ARIZONA

Rocks & Minerals

A Field Guide to the Grand Canyon State

Bob Lynch & Dan R. Lynch

Adventure Publications
Cambridge, Minnesota

Dedication

To Nancy Lynch, wife of Bob and mother of Dan, for her continued support of our book projects.

And to Julie Kirsch, Dan's wife, for always being there for Dan.

Acknowledgments

Thanks to George Godas, George Robinson, Ph.D., Ken Flood, Mike Anderson, Pat McMahan, Ben Grosz, Michael Shannon and Mitchell Dale for providing information and specimens.

Photography by Dan R. Lynch

Cover and book design by Jonathan Norberg

Edited by Brett Ortler

15 14 13 12 11 10

Arizona Rocks & Minerals
Copyright © 2010 by Bob and Dan R. Lynch
Published by Adventure Publications
An imprint of AdventureKEEN
310 Garfield Street South
Cambridge, Minnesota 55008
(800) 678-7006
www.adventurepublications.net
All rights reserved
Printed in China
ISBN 978-1-59193-237-6 (pbk.)

Table of Contents

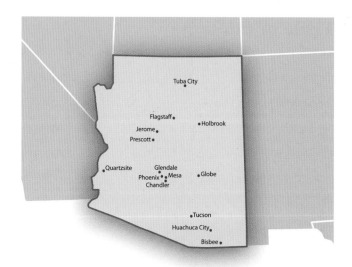

Introduction

Few states have more collectible minerals available than Arizona, and this is true even for the casual collector. With the high-altitude, wooded Colorado Plateau in the north, the low-lying deserts in the south and the rocky peaks in between, the Grand Canyon State has a more diverse topography than much of the country and offers a huge range of environments for rock hunting. The land is rich with metals, primarily copper, but it also contains iron and silver; all of these minerals react to weathering, creating new minerals. As a result, copper and silver mining have played a huge role in Arizona's history and many rare minerals are available to the public in museums and rock shops. So whether you're on a hillside, in a dried-up riverbed, or in one of the state's many rock shops, you're sure to find a treasured addition to your collection.

Rocks vs. Minerals

Many people go hunting for rocks and minerals without knowing the difference between the two. The difference is simple: a mineral is a crystallization of a chemical compound; for example, silicon dioxide crystallizes to form quartz, the most abundant mineral on earth. A rock is a mixture of many different minerals. While minerals exhibit very definite characteristics, rocks do not and vary greatly because they consist of a variety of minerals. This can make identification of rocks more difficult.

Protected Places

It is your responsibility to know where you can and cannot collect rocks and minerals in Arizona. Arizona has extensive national and state forests and deserts, as well as Native American reservations, where it is illegal to collect anything. In addition, Arizona's rich Native American history means that you'll likely come across ancient artifacts, such as arrowheads or pieces of pottery. Please note that it is illegal to remove these specimens from any site!

Stay Out of Abandoned Mines

Arizona has long been an active mining state, which means that hundreds of mines, new and old, dot the landscape. While many are privately owned and closed to the public, many more are simply abandoned and it is entirely possible that you may stumble across one in your travels. If you ever come across an unrestricted mine, don't go inside! As enticing as it may be to see the inside of an old mine shaft, they are extremely dangerous and unpredictable. Many are long abandoned and in disrepair and any human interaction could cause them to collapse. In addition, many contain high levels of dangerous gases and suffocation is a possibility.

The Desert Sun

The desert also holds many natural dangers. Inexperienced collectors who travel far into the dry climate looking for treasures often forget about Arizona's high temperatures and hard sunlight. Always be sure to wear plenty of sunblock, no matter how sheltered you think you may be, and wear a hat. Sunglasses will also protect your eyes and help keep you from squinting in the sunlight. And, most importantly, remember to bring plenty of water and to keep yourself well-hydrated.

Flash Floods

When it comes to desert weather, you have to worry about more than just the heat. Rainstorms, while infrequent, can be swift and violent. The dry desert surfaces can cause a downpour to quickly turn to a flood, especially in or near the many washes (dry river beds) found in desert areas. If you're out collecting when the sky is threatening rain, you need to seek out higher ground immediately.

Snakes, Spiders and Lizards

The Arizona climate is also home to potentially hazardous animals. Snakes, spiders and lizards all have their own defenses and should be avoided. Never attempt to approach, catch or scare away any animal, as you will probably be the one who ends up hurt. Luckily most of these animals come out only at night when the temperatures have dropped significantly. And in the winter months, many of these animals hibernate.

Also, when collecting, it always pays to watch your step. Most importantly, do not investigate holes in rocks or in the ground, as it might be another creature's home. Finally, turning over rocks may expose a rare gem, but may also reveal a coiled snake or an angry insect.

Cacti

Animals aren't the only Arizona lifeforms that have developed defenses. Collectors need to be as mindful of plant life, especially cacti, as they are of animal life. All too often, rockhounds walk through the desert with their eyes on the ground, instead of on the cactus they're about to walk into.

It is also important to note that the saguaro cactus, the largest species of cacti in Arizona, is protected, and harming one is illegal. Even collecting pieces of dead saguaros can get you a fine.

 Potentially Hazardous Rocks and Minerals

While the vast majority of Arizona's minerals are safe to handle and collect, a few have their own dangers. Potentially hazardous minerals in this book are identified with the symbol shown above; the hazards associated with them are listed here and are discussed in detail in the "notes" section on the pages listed below. Always take proper precautions when handling such materials; see the pages listed for specific precautions and advice.

Chalcanthite (page 67)—poisonous when ingested

Cinnabar (page 81)—contains mercury

Galena (page 119)—contains lead

Mountain Leather (page 169)—asbestos, can cause cancer

Serpentine (page 201)—some varieties are asbestos, which can cause cancer

Torbernite (page 217)—radioactive

Tremolite (page 221)—asbestos, can cause cancer

Glossary Note

We're aware that books about geology, rocks and minerals can be very technical. To help make this book intuitive for the lay-person but useful for the expert, we've included technical terms in the text, but we "translate" the technical phrases immediately after using them by providing a brief definition. Of course, all of the geology-related terms we've used are also defined in the glossary found in the back of this book.

Hardness and Streak

Anyone who wants to identify minerals needs to know about two techniques that can help identify a specimen—hardness and streak tests. All minerals yield results in both tests, and so will many rocks, and both techniques are indispensable in identifying your specimens.

The measure of how resistant a mineral is to abrasion is called hardness. The most common hardness scale, called the Mohs hardness scale, ranges from 1 to 10, with 10 being the hardest. An example of a mineral with a hardness of 1 is talc; it is a chalky mineral that you can easily scratch with your fingernail. A mineral with a hardness of 10 is diamond, which is the hardest naturally-occurring substance on earth and will scratch every-thing. Most minerals, Arizona's included, fall somewhere in the range of 2 to 7 on the Mohs hardness scale, so learning how to perform a hardness test is a key skill to have. Common tools used in a hardness test (also known as a scratch test) include your fingernails, a copper coin, a piece of glass, a steel nail and a pocket knife. There are also hardness test kits available that contain a tool of each hardness. On the following page is a chart that shows minerals of each hardness, as well as the hardnesses of common tools.

The second test every amateur geologist and rock collector should know is streak. When a mineral is crushed or powdered, it will have a distinct color—this color is the same as the streak color. When a mineral softer than a streak plate is rubbed along such a plate, it will leave behind a powdery stripe of color, called the streak. This is an important test to perform because sometimes the streak color will differ greatly from the mineral itself. Hematite, for example, is a dark, metallic and gray mineral, yet its streak is a rusty red color. Streak plates are sold in some rock and mineral shops, but if you cannot find one, a simple unglazed piece of porcelain will work. There are only two things you need to remember about streak tests: If the mineral is harder than the streak plate, it will not produce a streak and will instead scratch the streak plate. Secondly, never test rocks for streak, since they are made up of many different minerals.

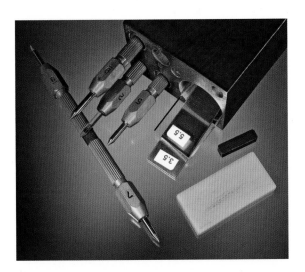

The Mohs Hardness Scale

The Mohs hardness scale is the primary measure of mineral hardness. This scale ranges from 1 to 10, from softest to hardest. Ten minerals commonly associated with the scale are listed here. Some common tools used to determine a mineral's hardness are listed here as well. If a mineral is scratched by a tool, you know it is softer than that tool's hardness.

HARDNESS	EXAMPLE MINERAL	TOOL
1	Talc	
2	Gypsum	
2.5		Fingernail
3	Calcite	
3.5		Copper Coin
4	Fluorite	
5	Apatite	
5.5		Glass, Steel Nail
6	Orthoclase	
6.5		Streak Plate
7	Quartz	
8	Topaz	
9	Corundum	
10	Diamond	

For example, if a mineral is scratched by a copper coin but not your fingernail, you can conclude that its hardness is 3, equal to that of calcite. If a mineral is harder than 6.5, or the hardness of a streak plate, it will have no streak unless weathered or altered by other, softer minerals, and will instead scratch the streak plate itself.

Quick Identification Guide

Use this quick identification guide to help you determine which rock or mineral you may have found. We've listed the primary color groups and some basic characteristics of the rocks and minerals of Arizona, as well as the page number where you can read more about your possible find. The most common traits for each rock or mineral are listed here, but be aware that your specimen may differ greatly.

	If white or colorless and...	then try...
	Soft, massive material with streaks of red throughout	alunite, page 37
	Small, soft crystals growing with galena	anglesite, page 41
	Delicate, branching crystals that closely resembles calcite	aragonite, page 45
	Thick bladed crystals that feel heavy for their size	barite, page 55
	Soft, abundant crystals or masses within other rock	calcite, page 63
	Bright, lustrous crystals growing at 60 degree angles to each other	cerussite, page 65
	Hard, waxy-surfaced masses often appearing lumpy	chalcedony, page 69

WHITE

Quick Identification Guide

(continued) **If white or colorless and...**	**then try...**
Massive, soft and glass-like	cristobalite, page 91
Formed in groups of small, square crystals with curved faces	dolomite, page 103
Extremely abundant, somewhat hard minerals, often found within granite	feldspar, page 109
Very soft, chalky crystals that are easily scratched with your fingernail	gypsum, page 133
Very abundant, soft rock that will fizz in vinegar	limestone, page 147
Porcelain-like masses forming veins or cauliflower-like pockets in rock	magnesite, page 151
Soft, massive and sometimes banded rock that can be scratched with a copper coin	marble, page 157
Soapy-feeling, clay-like mineral that swells when wet	montmorillonite, page 167
Very hard, glassy crystals or pockets in rock, especially granite	quartz, page 181

Quick Identification Guide

(continued)	**If white or colorless and...**	**then try...**
	Hard, grainy rock with a sandy texture	quartzite, page 183
	Extremely soft mineral that your fingernail will scratch	talc, page 213
	Mineral with radiating, fibrous crystals that sometimes feel silky	tremolite, page 221
	Fibrous crystals found within cavities in basalt and can appear "fuzzy"	zeolites, page 233

	If gray and...	**then try...**
	Dark, metallic mineral most often found growing on top of silver	acanthite, page 29
	Extremely abundant, dark rock found virtually everywhere	basalt, page 57
	Dark, soft, metallic mineral found with copper	chalcocite, page 71
	Massive, waxy-textured mineral	chert, page 75

Quick Identification Guide

(continued) **If gray and...**	**then try...**
Dark, tree-like growths on the surface of rocks, specifically limestone	dendrites, page 95
Dark rock with lighter spots of round feldspar crystals	diabase, page 97
Dark rock with shiny, coarse crystals within it	gabbro, page 117
Extremely heavy, metallic mineral with cubic crystals	galena, page 119
Fine, thin crystals often arranged into botryoidal (grape-like) crusts	hemimorphite, page 137
Heavy, metallic chunks, often coated in a layer of rust or rock	meteorites, page 159
Brightly lustrous, metallic crystals with a six-sided shape and bluish tint	molybdenite, page 165
Bright gray, shiny metal with a dark gray surface tarnish that can be scratched off	silver, page 207
Light rock that is grainy to the touch and can have fragments of other rocks within it	tuff, page 223

GRAY

Quick Identification Guide

GRAY

(continued)	**If gray and...**	**then try...**
	Flat, flaky crystals that are very flexible and reflective	mica, page 161
	Glassy, sharp rock full of small bubbles resembling froth	rhyolite, page 185

BLUE

	If blue and...	**then try...**
	Delicate, needle-like crystals arranged into rounded balls	aurichalcite, page 51
	Richly colored crystals that resemble rounded blades	azurite, page 53
	Long, fibrous, wavy crystals that are very brittle	chalcanthite, page 67
	Soft, vibrantly colored mineral found with other copper minerals	chrysocolla, page 79
	Darkly colored, very hard mineral found with quartzite	dumortierite, page 105
	Small, thin crystals arranged into botryoidal (grape-like) crusts	hemimorphite, page 137

Quick Identification Guide

(continued) **If blue and...**		**then try...**
	Tiny pointed crystals, often coated with a layer of a different, colorless mineral	kinoite, page 143
	Streaks throughout a lighter-colored rock that have striated (grooved) sides	kyanite, page 145
	Tiny crystals arranged into balls that appear soft and "fuzzy"	rosasite, page 187
	Hard, sky-blue mineral found within cavities in rock	turquoise, page 225

If black and...		**then try...**
	Stubby crystals embedded within rock	augite, page 49
	Botryoidal (grape-like) metallic mineral with a fibrous cross-section	goethite, page 127
	Botryoidal (grape-like) metallic mineral, often with a reddish coating	hematite, page 135
	Metallic mineral that a magnet will stick to	magnetite, page 153

Quick Identification Guide

BLACK

	(continued) **If black and...**	**then try...**
	Glassy rock that has a very sharp edge when broken	obsidian, page 171
	Sooty, metallic, fibrous mineral that can blacken your fingers	pyrolusite, page 179
	Very lustrous, pyramid-shaped crystals with a greenish tint	sphalerite, page 211
	Dull, dark mineral often mixed within malachite or other copper minerals	tenorite, page 215
	Long, thin crystals with striated (grooved) faces	tourmaline, page 219

YELLOW

	If yellow and...	**then try...**
	Metallic, brittle mineral that is found as cubic crystals	pyrite, page 177
	Gritty, abundant rock that appears to be made of sand	sandstone, page 191
	Small, square, glassy crystals growing on top of rock	wulfenite, page 231

Quick Identification Guide

BROWN

	If brown and...	then try...
	Hard, glassy crystals or masses, often embedded within hard rock	andalusite, page 39
	Dark, hard and unusually round rocks	concretions, page 83
	Long, thin, hollow tubes of sand that resemble tree branches	fulgurite, page 115
	Hard, round crystals with many faces that are embedded in rock	garnet, page 121
	Light, soft, clay-like material	kaolinite, page 141
	Very abundant, soft rock that will fizz in vinegar	limestone, page 147
	Rust-colored rock with no apparent crystal structure	limonite, page 149
	Soft, light, clay-like material that swells greatly when wet	bentonite, page 167
	Fibrous mats of material that appear as if woven together from fabric	mountain leather, page 169

Quick Identification Guide

(continued)

BROWN

If brown and...	then try...
Soft, "muddy" mineral coating the inside of cavities in rock	saponite, page 193
Glassy, square, metallic crystals	siderite, page 205

ORANGE

If orange and...	then try...
Small, needle-like crystals	mimetite, page 163
Small, rare, pyramid-shaped crystals	scheelite, page 195
Small, six-sided crystals that coat the surface of another rock or mineral	vanadinite, page 227
Square, glassy crystals that grow on the surface of other rocks and minerals	wulfenite, page 231

GREEN

If green and...	then try...
Six-sided needles that grow within other minerals or rocks	apatite, page 43

19

Quick Identification Guide

(continued) **If green and...** | **then try...**

GREEN

	Coatings of dull, small, needle-like crystals	atacamite, page 47
	Glassy, emerald-colored, needle-like crystals arranged into circular groupings	brochantite, page 61
	Small spots of greenish color within rocks like basalt	chlorite, page 77
	Tiny grass-green balls growing on the surface of another rock or mineral	conichalcite, page 87
	Green streaks within a softer, whiter rock, such as marble	diopside, page 99
	Tiny, flat, teal-colored crystals growing on the surface of another rock or mineral	dioptase, page 101
	Striated (grooved) yellow-green crystals or pockets within rock	epidote, page 107
	Glassy crystal masses often found growing in bands of color	fluorite, page 111
	Botryoidal (grape-like) masses with a banded cross-section	malachite, page 155

Quick Identification Guide

GREEN

	(continued) **If green and...**	**then try...**
	Thin, flat, bladed crystals growing together in book-like groupings	mica, page 161
	Clusters of small green crystals clumped together within basalt	olivine, page 173
	Greasy- or silky-feeling green minerals	serpentine, page 201
	Soft, layered rock that easily pulls apart in thin, flat sheets	shale, page 203
	Botryoidal (grape-like) coatings of a glassy mineral within cavities in rock	smithsonite, page 209
	Dark, lustrous crystals shaped like two pyramids end-to-end	sphalerite, page 211
	Tiny, square crystals	torbernite, page 217
	Hard, dark fibrous masses	actinolite, page 221
	Tiny, rare, glassy six-sided crystals	willemite, page 229

Quick Identification Guide

PINK OR VIOLET

	If pink or violet and...	then try...
	Hard, waxy mineral, often with a white to brown outer coating	purple agate, page 35
	Extremely common mineral, most often found in granite	feldspar, page 109
	Glassy, soft crystals or masses, often found growing in bands of color	fluorite, page 111
	Fibrous crystals found growing in cavities within basalt	zeolites, page 233

RED

	If red and...	then try...
	Tiny veins of bright red or pink within quartz or other material	cinnabar, page 81
	Cubic crystals with a high luster that appear almost metallic	cuprite, page 93
	Very hard, round crystals found embedded in rock	garnet, page 121
	Hard, waxy masses of opaque material	jasper, page 139

Quick Identification Guide

RED

(continued) **If red and...**	**then try...**
Abundant, hard, red rock that often has parallel bands	rhyolite, page 185
Stubby, flat crystals embedded within another mineral	rutile, page 189
Abundant, gritty rock that appears as if made of sand	sandstone, page 191
Layered, soft rock that is found in large, flat beds with cream-colored banding	catlinite, page 203
Small, glassy six-sided crystals that grow on the surface of another rock or mineral	vanadinite, page 227

METALLIC

If metallic and...	**then try...**
Gray coating on the surface of silver	acanthite, page 29
Brassy metallic mineral with a fast-forming multi-colored surface tarnish	bornite, page 59
Gray, shiny, abundant, metallic mineral that is very soft	chalcocite, page 71

Quick Identification Guide

(continued) **If metallic and...**	**then try...**
Brittle, gold-colored metallic mineral with a multi-colored tarnish	chalcopyrite, page 73
Reddish metal, sometimes with a green or black surface tarnish	copper, page 89
Very heavy, cubic mineral	galena, page 119
Botryoidal (grape-like) black masses with a fibrous cross-section	goethite, page 127
Bright yellow metal that is bent easily	gold, page 129
Botryoidal (grape-like) gray masses that often have a reddish surface coating	hematite, page 135
Black, metallic, pyramid-shaped crystals that a magnet will stick to	magnetite, page 153
Heavy, dark chunks, often with a rusty coating	meteorites, page 159
Bluish gray, layered, six-sided crystals that are very soft and bend easily	molybdenite, page 165

METALLIC

Quick Identification Guide

	If metallic and...	**then try...**
	Brittle, hard, yellow mineral that often forms into cubes	pyrite, page 177
	Sooty, fibrous, black mineral that can blacken your hands	pyrolusite, page 179
	Bright gray metal, often with a dark gray, dull coating	silver, page 207

METALLIC

	If multi-colored or banded and...	**then try...**
	Rounded balls of hard, banded rock within cavities in basalt	agate, page 31
	Hard, waxy lumps that exhibit flashes of color when rotated	fire agate, page 33
	Brassy metallic mineral that forms a multi-colored tarnish very quickly on a fresh break	bornite, page 59
	Massive, hard, waxy mineral	chalcedony, page 69
	Brittle, gold-colored, metallic mineral with a multi-colored tarnish	chalcopyrite, page 73

MULTI-COLORED OR BANDED

25

Quick Identification Guide

(continued)	If multi-colored or banded and... then try...
	Rock that appears to be conglomerate or formed of smaller rocks or rock fragments — breccia, page 85
	Rocks or minerals that resemble living things, such as trees or coral — fossils, page 113
	Round balls of rock that are hollow and can be broken to reveal crystals — geodes, page 123
	Coarsely banded or layered rocks, such as granite — gneiss, page 125
	Light-colored rocks with many coarse grains of light- and dark-colored minerals — granite, page 131
	Hard, waxy, massive mineral that often has multi-colored bands — jasper, page 139
	Masses of parallel-banded minerals — onyx, page 175
	Finely-layered rocks, often with a glittery appearance — schist, page 197

Sample Page

HARDNESS: 7 **STREAK:** White

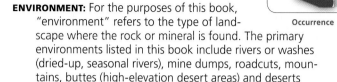

Occurrence

ENVIRONMENT: For the purposes of this book, "environment" refers to the type of landscape where the rock or mineral is found. The primary environments listed in this book include rivers or washes (dried-up, seasonal rivers), mine dumps, roadcuts, mountains, buttes (high-elevation desert areas) and deserts

WHAT TO LOOK FOR: Common or characteristic traits of the rock or mineral

SIZE: The general size range of a rock or mineral

COLOR: The general colors a rock or mineral exhibits in its natural state

OCCURRENCE: How easy or difficult this rock or mineral is to find. "Very common" means the material takes virtually no effort to find. "Common" means the material can be found with little effort. "Uncommon" means the material may take a good deal of hunting to find. "Rare" means the material will take great lengths of time and much energy to find. "Very rare" means the material is so uncommon that you will be lucky to even find a trace of it

NOTES: Additional information about the rock or mineral, including more extensive advice about what to look for, specific methods to differentiate it from other rocks and minerals, and interesting facts or unique characteristics

WHERE TO LOOK: Specific areas or locations within the state that are good places to start looking for the rock or mineral. This section often refers to particular mountain ranges and mining districts

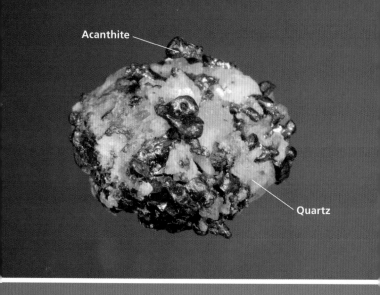

Acanthite

HARDNESS: 2–2.5 **STREAK:** Metallic black

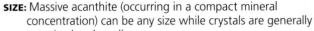

ENVIRONMENT: Mine dumps, mountains

WHAT TO LOOK FOR: Dark gray, soft metal crystals or as tarnish on silver specimens

Occurrence

SIZE: Massive acanthite (occurring in a compact mineral concentration) can be any size while crystals are generally pea-sized and smaller

COLOR: Dark lead-gray

OCCURRENCE: Very rare

NOTES: Acanthite is a combination of silver and sulfur that has long been mined as a silver ore. It can be told apart from silver primarily by its darker gray color but also by its slightly softer hardness. Acanthite occurs most often on silver specimens; in fact, if you have silver jewelry, you might be able to find acanthite in your own home—the dark gray tarnish that forms on silver is actually acanthite. In fact, most natural specimens of acanthite are found as a coating of dark tarnish on silver specimens, and well-formed crystals of acanthite are very rare.

Acanthite is an interesting mineral because its crystal structure changes once it reaches a temperature of about 344° F. This high-temperature form is called argentite and it has a completely different internal crystal structure. Once the specimen is allowed to return to room temperature, its internal structure once again becomes that of acanthite. Acanthite is often found with quartz, barite, bornite, pyrite and, of course, silver.

WHERE TO LOOK: Look in the hills west of Tombstone for silver and in the mine dumps in southeast Arizona, especially near Globe.

Arizona agates

Rough

Polished

Agates from Fourth of July Butte

Cut

Rough

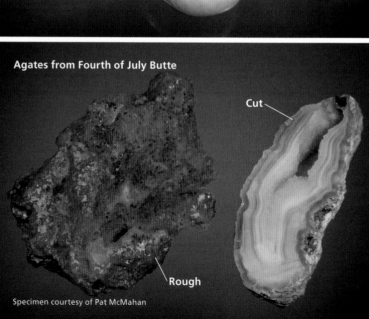

Specimen courtesy of Pat McMahan

Agate, Common Varieties

HARDNESS: 7 **STREAK:** White

Occurrence

ENVIRONMENT: Washes, rivers, buttes, roadcuts

WHAT TO LOOK FOR: Translucent nodules (balls) of material that show ring-like banding inside

SIZE: Agates are generally no larger than your fist, though some can be quite large

COLOR: White to gray, red, brown, blue to purple

OCCURRENCE: Uncommon

NOTES: Agates are a form of chalcedony, which, like chert and jasper, is a variety of microcrystalline quartz (quartz crystals too small to be seen with the naked eye). In chalcedony, the microscopic quartz crystals arrange themselves into tiny parallel fibers that make the mineral somewhat translucent. However, whereas chalcedony forms in very large masses, agates form within the vesicles (gas bubbles) in basalt. And while no one knows exactly how the different bands of color actually form, it is thought that the chalcedony formations somehow arrange themselves into concentric rings (bull's-eye pattern). It is helpful to think of each band as a chalcedony coating, or shell, around smaller shells of chalcedony. Chalcedony is white or colorless when pure, but it is very commonly stained by other minerals. Iron oxides, like hematite or goethite, frequently turn agates red, brown or yellow.

Arizona is home to many unique varieties of agate. "Arizona agates," as they are sometimes called, are simple, white agates formed of nearly pure chalcedony and have little to no variation in color. Agates from the Fourth of July Butte, near Phoenix, are known for their white and bluish bands.

WHERE TO LOOK: Look for agates anywhere there is exposed basalt.

Fire agate

Rough

Polished

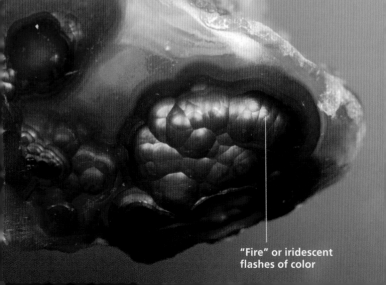

"Fire" or iridescent flashes of color

Agate, Fire

HARDNESS: 7 **STREAK:** White

Occurrence

ENVIRONMENT: Desert, mountains

WHAT TO LOOK FOR: Brown, lumpy, translucent material that exhibits brilliant flashes of color when broken or polished

SIZE: Fire agates are normally no larger than your fist

COLOR: Brown to red, white to gray, with colorful iridescence

OCCURRENCE: Rare

NOTES: While not an agate in the traditional sense, fire agates are unique, collectible stones that are well-known in Arizona. Most do not exhibit the characteristic concentric (bull's-eye pattern) banding of agates and instead show an internal, opal-like iridescence and flashes of color. It's this "fire" that makes fire agates some of Arizona's most sought-after specimens.

Like most agates, the brown or reddish coloration is due to the presence of iron oxides in the material. And though most agates are found as nodules (round masses that form within rock), fire agates form in open pockets of rock from a dripping solution of silica (quartz material). Each newly deposited layer of silica builds upon the layers beneath it, and the result is a lumpy, rounded mass with a very distinct look. Each of these layers of silica is very thin and contains different amounts of iron oxide, which makes some layers clearer than others. This combination causes light to bounce between the layers of the agate, creating the agate's famous "fire."

WHERE TO LOOK: Southeast of Kingman about 20 miles in the Hualapai Mountains area, specifically the northeast side of the hills, as well as in the hills between Kingman and the California border.

Purple agate from Burro Creek, in the Bagdad area

Rough

Polished

Purple agate (cut) from Sheep Bridge on the Verde River

Specimen courtesy of Pat McMahan

Agate, Purple

HARDNESS: 7 **STREAK:** White

Occurrence

ENVIRONMENT: Rivers, washes, buttes, mountains

WHAT TO LOOK FOR: Hard purple material, often with a lighter-colored outer coating

SIZE: These agates generally occur softball-sized and smaller

COLOR: Purple to blue internally; white to gray and yellow to brown externally

OCCURRENCE: Rare

NOTES: Agates normally occur in shades of brown, red or white, but Arizona is home to several locations that produce rare purple agates. While no one knows exactly what causes this strange coloration, purple agates presumably form because of the same process that creates amethyst (purple quartz). Amethyst consists of quartz that includes iron and aluminum impurities. Aluminum tends to turn clear quartz to shades of gray, but when iron is present along with the aluminum, the combination results in beautiful shades of violet.

Two of the primary locations for purple agates in Arizona are Burro Creek and Sheep Bridge and each produces different types of agates. Agates from Burro Creek, near Bagdad, Arizona, don't exhibit much banding, and many specimens would be more appropriately labelled "purple chalcedony." Sheep Bridge, at Sheep Crossing on the Verde River, northeast of Phoenix, produces purple agates that often contain needle-like structures, called sagenite.

WHERE TO LOOK: Burro Creek is just north of Bagdad, which is southeast of Kingman. Sheep Crossing is northeast of Phoenix on the Verde River, which runs along the eastern side of Phoenix.

Rough alunite

Polished specimens

Alunite

HARDNESS: 3.5–4   **STREAK:** White

ENVIRONMENT: Desert, buttes

WHAT TO LOOK FOR: Light-colored rock, often with streaks of red or pink

SIZE: Alunite normally occurs massively and can be found in any size

COLOR: White to gray, often with streaks or patches of red or pink

OCCURRENCE: Common

NOTES: Alunite is the result of rocks rich in feldspar being altered by acidic water. Rhyolite is the primary type of rock affected by this process, and sometimes enormous, mountain-sized rhyolite formations are transformed into alunite.

Alunite is generally massive (a compact mineral concentration) and can range in size from pebbles to boulders. Very rarely, alunite crystals can form within cavities in massive alunite formations. Alunite's primary colors are white or cream-colored to gray and it often has red or flesh-colored streaks or patches throughout. Specimens without the reddish coloration can be easily confused with limestone. A simple test to tell the two apart is to put strong vinegar on each of the materials. A specimen of limestone will effervesce and fizz in the acid but alunite will not.

Alunite can be polished to produce attractive specimens for collectors and it is often cut into cabochons (shaped, polished gemstones) for use in jewelry, but it is not a very durable stone due to its low hardness.

WHERE TO LOOK: The hills about 5 miles west of Quartzsite are known for alunite, as well as the Patagonia Mountains in Santa Cruz county, northeast of Nogales.

Massive andalusite

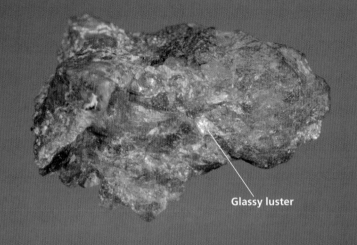

Glassy luster

Chiastolite

Polished specimens

"Four-leaf clover" cross-section

Andalusite

HARDNESS: 7.5 **STREAK:** Colorless

ENVIRONMENT: Mountains, buttes, desert, mine dumps

Occurrence

WHAT TO LOOK FOR: Coarse, square-sided crystals, often embedded within another rock

SIZE: Crystals are generally short and stubby, no more than a few inches, while massive andalusite (a solid, compact mineral concentration) can occur in a wide range of sizes

COLOR: Reddish brown to gray

OCCURRENCE: Uncommon

NOTES: Andalusite is a hard, aluminum-rich mineral that often occurs as rough, stubby crystals embedded in rock. Andalusite crystals are often described as "cigar-like" because of their blunt ends and poorly-formed, round crystal faces. Andalusite is found in well-crystallized granite formations (pegmatites) and in metamorphic rocks (rocks formed under pressure), which is similar to the way garnet develops.

Andalusite generally isn't much to look at and specimens are brown or reddish with a slightly glassy luster. Andalusite crystals are more desirable but less common; nevertheless, a variety of andalusite known as chiastolite is the exception to this rule. Good examples of chiastolite exhibit a cross-section that resembles a checkerboard or a four-leaf clover. This is because when the andalusite was forming, its impurities were forced into the corners and center of the crystals. This results in attractive specimens that are often used in jewelry. Andalusite is often found with quartz, feldspars and micas.

WHERE TO LOOK: The Bradshaw Mountains south of Prescott and the Chiricahua Mountains northeast of Douglas.

Anglesite crystals

Galena

Massive anglesite bands

Anglesite

HARDNESS: 2.5–3 **STREAK:** White to gray

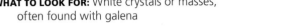

Occurrence

ENVIRONMENT: Mine dumps, mountains

WHAT TO LOOK FOR: White crystals or masses, often found with galena

SIZE: Crystals are generally thumbnail-sized or smaller while massive anglesite (a compact mineral concentration) can be any size.

COLOR: Colorless, white or gray

OCCURRENCE: Rare

NOTES: Anglesite is a rare lead-based mineral that forms primarily as delicate, glassy crystals, though it also can be found as massive pieces that are sometimes banded with cerussite and galena. It forms in weathering deposits of lead and is often found with galena, the primary ore of lead, sometimes growing right on the surface of a galena crystal. It is less common than its lead-based cousin, cerussite, which also can occur on galena, but is slightly harder than anglesite. Anglesite's hardness and high luster, as well as its association with other lead minerals, are normally enough to identify it. As mentioned above, you're only likely to confuse it with cerussite, though anglesite is notably softer. And massive anglesite obviously will not have any observable crystal structure; you may have to rely solely on color, hardness, and surrounding minerals to identify it. Some of the best anglesite crystals are found with massive or granular galena rather than fine crystals of galena. This is because the massive varieties allow more mineral solutions to permeate through them, whereas crystals do not. This results in better formed lead-based mineral crystals growing with the galena.

WHERE TO LOOK: Mine dumps near Bisbee and Tombstone are good places to start.

Apatite crystals

Calcite

Calcite

Green, needle-like apatite crystals

Apatite

HARDNESS: 5 **STREAK:** White

ENVIRONMENT: Mine dumps, buttes, mountains

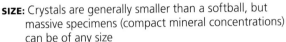

Occurrence

WHAT TO LOOK FOR: Light-colored crystals or masses with a greasy or glassy luster

SIZE: Crystals are generally smaller than a softball, but massive specimens (compact mineral concentrations) can be of any size

COLOR: Colorless, white to gray and green are most common, but apatite can also be yellow, blue or pink

OCCURRENCE: Common

NOTES: Apatite is a fairly common mineral that forms in a wide variety of mineral environments, though primarily within rocks. Characteristic six-sided, elongated crystals are the most desirable form, but apatite can also occur massively. While green and white are the most common colors, apatite can also occur in blue, yellow and pink hues, but well-formed blue crystals are the most sought after. No matter what the color, apatite always gives a white streak, making even the most vivid samples easier to identify. In Arizona, apatite is quite common, but not as loose, well-formed crystals. As mentioned above, a lot of apatite is incorporated into rocks, like granite, and unless you're looking for it, you're not likely to easily find it. Other times, thin needles of apatite will be encased within another mineral, such as calcite. Apatite is the most common phosphorus-bearing mineral and as such it is the primary source for plants to absorb the important element. Another interesting fact is that apatite has nearly the same structure and chemical composition as bone.

WHERE TO LOOK: The Aquarius Mountains south of Kingman and the Bradshaw Mountains south of Prescott.

Delicate, branching crystals

Aragonite replacement of a glauberite crystal

Aragonite

HARDNESS: 3.5–4   **STREAK:** White

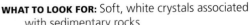

ENVIRONMENT: Mine dumps, mountains, buttes

WHAT TO LOOK FOR: Soft, white crystals associated with sedimentary rocks

Occurrence

SIZE: Crystals can be virtually of any size but are generally smaller than a basketball

COLOR: Colorless, white to gray and brown or yellow depending on impurities

OCCURRENCE: Common

NOTES: Aragonite is a common mineral with essentially the same chemical composition as the much more common calcite. The two are distinguished primarily by crystal structure, as aragonite occurs in many different crystal habits (forms) than calcite; for instance, aragonite sometimes occurs in complex branching crystals, whereas calcite does not. A hardness test can also differentiate the two, as aragonite is slightly harder than calcite. If you are having a hard time telling aragonite apart from calcite, it is safe to assume that your specimen is calcite, since aragonite is considerably rarer. Both minerals are generally colorless or white; this makes color a poor method of distinguishing the two.

Since calcite and aragonite are so closely related, they are often found together in the same specimen. Aragonite is also often found alongside other minerals' crystals and even replaces some of them. Aragonite occurs in the animal kingdom too—mollusks and other aquatic animals actually produce their own aragonite, the iridescent interior of the oyster's shell (better known as mother-of-pearl) is perhaps the best example.

WHERE TO LOOK: The Dragoon Mountains near Gleeson as well as the mine dumps and hills near Tombstone.

Atacamite coating

Azurite crystal

Limonite matrix

Needle-like crystals

Atacamite

HARDNESS: 3–3.5 **STREAK:** Green

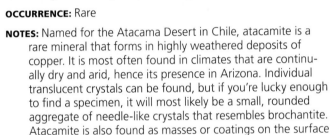

Occurrence

ENVIRONMENT: Mine dumps

WHAT TO LOOK FOR: Tiny, green needle-like crystals or coatings

SIZE: Individual crystals are just millimeters in length

COLOR: Light to dark green, emerald green

OCCURRENCE: Rare

NOTES: Named for the Atacama Desert in Chile, atacamite is a rare mineral that forms in highly weathered deposits of copper. It is most often found in climates that are continually dry and arid, hence its presence in Arizona. Individual translucent crystals can be found, but if you're lucky enough to find a specimen, it will most likely be a small, rounded aggregate of needle-like crystals that resembles brochantite. Atacamite is also found as masses or coatings on the surface of other minerals and rocks.

Atacamite can be hard to differentiate from malachite crystals or brochantite, as mentioned above. Since the crystals of each mineral are generally too small to perform scratch or streak tests, and their colors and crystal structure can be so similar, you may have to rely on instinct. Malachite is more common than brochantite, and both minerals are much more common than atacamite. If all else fails, you can probably assume that the specimen in question is malachite. In fact, atacamite can actually become malachite or chrysocolla when affected by other chemicals entering its composition.

WHERE TO LOOK: Many of southern Arizona's copper mines produced atacamite and any mine dumps in the many mountain ranges between Phoenix and Tucson are a good place to start.

Augite crystals

Augite

HARDNESS: 5–6 **STREAK:** Green to gray

ENVIRONMENT: Mine dumps, mountains, roadcuts

Occurrence

WHAT TO LOOK FOR: Dark prismatic crystals
embedded within other rocks, such as granite

SIZE: Most crystals are small and generally no more than
an inch long

COLOR: Black to yellow-brown or green

OCCURRENCE: Common

NOTES: Augite is the most common member of the pyroxene
mineral group. Pyroxenes are dark-colored minerals that are
referred to as "rock-builders" because they are important
components in the formation of rocks like basalt, diorite and
gabbro. As such, augite is generally not recognizable on its
own. Instead it often occurs as dark spots or crystals in
other rock. It takes a strong microscope and a trained eye
to identify augite crystals within a dark piece of rock, like
gabbro. Individual augite crystals are easier to identify,
though rare; they grow freely in lighter-colored host rock.
In these situations, augite is found as small, short, dark
brown or black crystals and the crystal faces can often be
well-formed and attractive as collectible specimens.

Augite can also be found massively; such specimens are
slightly more difficult to identify. Thankfully, massive chunks
(solid, compact mineral concentrations) of augite break apart
at nearly ninety degree angles. This trait, combined with
hardness and streak color, is usually enough to identify it.

WHERE TO LOOK: Rocks rich in augite, like gabbro, can be found
around Flagstaff, but well-formed crystals can be found
in the hills near Tombstone, as well as in the Bradshaw
Mountains, south of Prescott, and the Patagonia Mountains,
northeast of Nogales.

Aurichalcite (blue)

Aurichalcite (blue) on hemimorphite (gray)

Specimen courtesy of George Godas

Aurichalcite

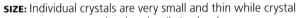

HARDNESS: 2 **STREAK:** Pale green

ENVIRONMENT: Mine dumps

WHAT TO LOOK FOR: Small, thin, blue crystals in radiating aggregates

Occurrence

SIZE: Individual crystals are very small and thin while crystal aggregates can be thumbnail-sized or larger

COLOR: Greenish blue

OCCURRENCE: Uncommon

NOTES: Many of the world's finest aurichalcite specimens have come from Arizona. Its beautiful blue-green color and delicate crystals make it very desirable for collectors. The tiny crystals are most often needle-like in shape and arranged in radiating aggregates, forming little "balls" of the mineral. Aurichalcite can also occur as crusts of thin, scaly crystals on a matrix.

Aurichalcite's color and crystal habit (form) are normally enough to identify it from other minerals. It can occur with azurite and malachite, which are also blue and green in color, but neither crystallize in quite the same way. Most of the time, aurichalcite crystals are too tiny and much too brittle to perform a hardness test, so it's often better to forgo this test when identifying a specimen.

Aurichalcite contains zinc and copper; in turn, both of these minerals are components of brass. It has therefore been said that aurichalcite is a "brass ore," but because aurichalcite is quite uncommon and found in such small quantities, commercial use isn't feasible.

WHERE TO LOOK: The mine dumps in the Mammoth area, northeast of Tucson, as well as the Patagonia Mountains, northeast of Nogales and the Dragoon Mountains, near Tucson.

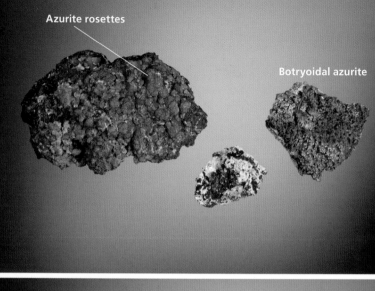

Azurite rosettes

Botryoidal azurite

Botryoidal azurite (blue) on malachite (green)

Bladed azurite crystals

Azurite rosette

Polished azurite cabochon with cuprite (red)
Specimen courtesy of Mitchell Dale

Azurite

HARDNESS: 3.5–4 **STREAK:** Light blue

ENVIRONMENT: Mine dumps

Occurrence

WHAT TO LOOK FOR: Deep blue crystals or masses, often occurring with green malachite

SIZE: Crystals are pea-sized and smaller, but crystal masses and massive azurite (compact mineral concentrations) can be much larger

COLOR: Rich, deep blue to light blue

OCCURRENCE: Uncommon

NOTES: Azurite is one of Arizona's most collectible and sought-after copper-related minerals. Named for its beautiful, rich azure-blue color, well-formed azurite crystals are prized specimens among collectors. Crystals range from drusy (crystal-covered) coatings to larger, tabular (flat) formations, as well as rosettes of thin, bladed crystals. Azurite can also form in botryoidal (grape-like) masses.

Azurite is very commonly found associated with malachite, as they are nearly identical in chemical composition, though azurite is rarer. The primary difference between the two minerals is that azurite contains one extra copper molecule. This makes azurite's chemical structure less stable than malachite's and after long periods of time azurite will actually turn into malachite. This contributes to the fact that the two minerals are very often found together in the same specimen. However, this relationship also makes azurite very easy to identify. Azurite occurs with malachite, limonite, chrysocolla and calcite, as well as other copper-related minerals, such as tenorite and cuprite.

WHERE TO LOOK: Found in many mines in southeast Arizona, especially near Bisbee, Tucson, Superior and Globe as well as many others in Cochise County.

Lamellar (gill-like) crystal grouping

Clear, bladed crystals

Poorly defined crystals

Barite

HARDNESS: 3–3.5 **STREAK:** White

Occurrence

ENVIRONMENT: Mine dumps, mountains, buttes, roadcuts

WHAT TO LOOK FOR: Tabular (flat) or blade-like crystals that feel very heavy for their size

SIZE: Barite specimens are generally smaller than a softball

COLOR: Colorless or white to gray, also yellow to brown and red depending on impurities

OCCURRENCE: Common

NOTES: Barite is a common mineral that is easily identified. Barite generally occurs as well-formed crystals within cavities in sedimentary rock. Since this means that barite's crystal structure is often very apparent, it is easy to determine the structure and shape of the mineral. These crystals are normally thin blades arranged in rosettes, called "desert roses," or lamellar (gill-like) formations, but they can also be thick and tabular (flat). Crystals can sometimes be quite thick and very large in size because of the mineral's abundance.

Aside from crystal structure, one of the best identifiers for barite is its high specific gravity. This means that the mineral feels very heavy for its size and even a small sample may surprise you with its weight. This is a rare trait for a light-colored mineral and is normally enough to tell it apart from a similarly colored sample of feldspar. Barite can occur with calcite, quartz and gypsum.

WHERE TO LOOK: Mine dumps and exposed rocks in the Pinal Mountains, east of Phoenix, the Big Horn Mountains, west of Phoenix, and the Santa Rita Mountains, north of Nogales, are good places to look.

Vesicle (gas bubble)

Basalt with vesicular minerals (minerals that fill gas bubbles in rock)

Basalt

HARDNESS: 5–6 **STREAK:** N/A

Occurrence

ENVIRONMENT: Mine dumps, buttes, mountains, roadcuts

WHAT TO LOOK FOR: Dark gray fine-grained rock

SIZE: Basalt can be found in any size, from pebbles to boulders

COLOR: Dark gray to black, sometimes brown

OCCURRENCE: Very common

NOTES: Basalt is a dark, compact rock that is the result of magma reaching the earth's surface and cooling very rapidly. Unlike granite, which cooled very slowly within the earth, allowing large crystals to form within it, basalt cooled rapidly, preventing visible crystals from developing. This dark, hard rock is not quite as common in Arizona as rhyolite, another volcanic rock, but it can be found virtually anywhere, nevertheless.

Basalt is often rich with vesicles, or bubbles, that were formed by gases trapped in the rapidly cooling rock. This is called vesicular basalt. These vesicles then capture other minerals that percolate through the rock as chemical solutions. As a result, the vesicles within basalt are home to quartz, agates, calcite and zeolites, as well as a host of other minerals. Some samples of vesicular basalt are so filled with cavities that it becomes unique and is known as scoria.

The primary minerals that make up basalt's dark appearance are plagioclase feldspar, olivine and augite, which is a variety of pyroxene. These minerals, however, are only observable in basalt under a very powerful microscope.

WHERE TO LOOK: Basalt can be found virtually anywhere, especially in the northern half of the state. It is much less common in sandy, desert areas, however.

Massive bornite

Multi-color tarnish

Granular bornite

Grainy surface texture

Bornite

HARDNESS: 3 **STREAK:** Grayish black

ENVIRONMENT: Mine dumps

WHAT TO LOOK FOR: Metallic, copper- or bronze-colored mineral, often with a very colorful surface coating

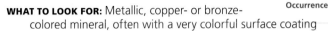

Occurrence

SIZE: Bornite occurs massively (in compact mineral concentrations) and can be found in any size

COLOR: Copper- to bronze-colored with a colorful tarnish containing blues, reds and purples

OCCURRENCE: Common

NOTES: Bornite, more commonly referred to as "peacock ore," due to its colorful and iridescent tarnish, is an ore of copper closely related to chalcopyrite and chalcocite, with which bornite can sometimes be confused. Chalcopyrite, in particular, is similar because of a nearly identical multi-colored tarnish. However, bornite has a key trait that chalcopyrite does not—upon breaking or scratching, it tarnishes quite rapidly to a bright purple color on the freshly exposed material. Because of their close relation to one another, bornite, chalcopyrite and chalcocite are often found in the same specimen. Bornite forms in compact masses, and while crystals of bornite do exist, they are extremely rare. Bornite also can be found as granular masses, which looks just like massive bornite with a grainy surface texture. Bornite's iridescent surface coloration makes it a desirable and showy mineral for collectors. Specimens available commercially are often sold as "peacock ore" and many of these samples are actually chalcopyrite, due to that mineral's availability.

WHERE TO LOOK: Mine dumps in the Bradshaw Mountain range, south of Prescott, and the Santa Catalina Mountains, near Tucson.

Chrysocolla (light blue)

Limonite matrix

Needle-like brochantite crystals

Tenorite (black)

Specimen courtesy of George Godas

Brochantite

HARDNESS: 3.5–4 **STREAK:** Pale green

ENVIRONMENT: Mine dumps

WHAT TO LOOK FOR: Very small, needle-like green crystals growing in radiating aggregates

Occurrence

SIZE: Individual crystals are mere millimeters long but crystal aggregates can be thumbnail-sized

COLOR: Emerald-green to dark green

OCCURRENCE: Uncommon

NOTES: A product of deeply altered copper deposits, brochantite is a beautiful, brilliantly colored variety of crystal that is desirable to collectors. Its tiny, needle-like crystals form in radiating aggregates that often fill cavities within the matrix (base rock). These crystals are generally very small and brittle and you should avoid touching them because they can break extremely easily.

Brochantite's emerald-green color and crystal growth habit (form) are normally enough to distinguish it. However, malachite sometimes forms similar small crystals, in which case chemical tests (which destroy part of the specimen) would need to be done to tell the two apart.

Brochantite can be found in many different rocks, though normally not limestone. Orange-brown limonite is a common matrix for crystals. Brochantite can also occur with chalcopyrite and cerussite crystals.

WHERE TO LOOK: The mine dumps in the Sierrita Mountains, southwest of Tucson, the Slate Mountains, south of Casa Grande, and the hills around Tombstone are good places to start.

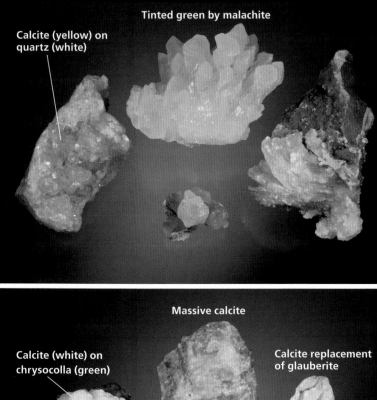

Tinted green by malachite

Calcite (yellow) on quartz (white)

Massive calcite

Calcite (white) on chrysocolla (green)

Calcite replacement of glauberite

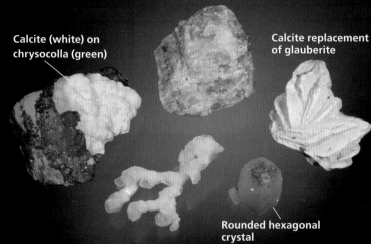

Rounded hexagonal crystal

Calcite

HARDNESS: 3 **STREAK:** White

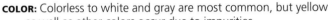

ENVIRONMENT: Found in all environments

WHAT TO LOOK FOR: Soft, light-colored crystals

Occurrence

SIZE: Calcite crystals can be pea-sized up to softball-sized and larger, while massive calcite (compact mineral concentrations) can be any size

COLOR: Colorless to white and gray are most common, but yellow as well as other colors occur due to impurities

OCCURRENCE: Very common

NOTES: One of the most common minerals on earth, calcite occurs in every mineral environment and has a huge number of crystal varieties. Calcite crystals in general, however, are hexagonal (six-sided) and prismatic. Calcite occurs with hundreds of other minerals and is quite easy to find no matter where you are looking. Most calcite specimens are colorless or white, though it can be found in many other colors, depending on other mineral impurities within it.

Quartz, a much harder mineral, is sometimes confused with calcite, but a hardness test is normally enough to determine a sample of calcite. Calcite also effervesces (bubbles) in acids and will dissolve away, leaving behind other minerals that may have been hidden within or underneath the calcite. Massive calcite is also common and is often banded. This variety is called calcite onyx and is widely used as a decorative stone. Calcite is also the primary constituent of various rocks, such as limestone, a sedimentary rock that often contains fossils, and marble, a metamorphic rock that is used as an architectural stone.

WHERE TO LOOK: Calcite is found everywhere, but the hills and mine dumps west of Tombstone as well as the mine dumps south of Wickenburg are good places to hunt.

Limonite matrix

Massive cerussite

Tiny cerussite crystals

Crust of poorly formed crystals

Cerussite

HARDNESS: 3–3.5 **STREAK:** White

ENVIRONMENT: Mine dumps

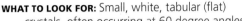

Occurrence

WHAT TO LOOK FOR: Small, white, tabular (flat) crystals, often occurring at 60 degree angles to each other, resembling latticework

SIZE: Crystals are generally thumbnail-sized and smaller, while massive cerussite (compact mineral concentrations) can occur in any size

COLOR: White to gray

OCCURRENCE: Uncommon

NOTES: Cerussite is a lead-based crystal that is often found associated directly with galena, the primary ore of lead. Cerussite's small, white, tabular (flat) crystals commonly form intricate lattice-like crystal networks, with individual crystals growing at 60 degree angles to one another. Cerussite crystals are generally white or gray and have a high specific gravity, meaning that they are very heavy for their size. This trait, combined with the mineral's high luster, is normally enough to distinguish cerussite from other similar looking minerals.

Cerussite can also occur massively, either in white or gray masses or in a banded form that includes other lead-based minerals like anglesite. Collectors, however, prefer to seek out crystal specimens of cerussite, many of which are found growing on or in limonite or limestone.

WHERE TO LOOK: You're likely to only find cerussite in mine dumps, especially those in the Huachuca Mountains, the Santa Rita Mountains and the Patagonia Mountains, all of which are northeast of Nogales.

Natural chalcanthite

Delicate, wavy blue crystals

Crumbly matrix

Man-made chalcanthite

Large, bladed crystals

Solid single-rock matrix

⚠️ Chalcanthite

HARDNESS: 2.5   **STREAK:** White

ENVIRONMENT: Mine dumps

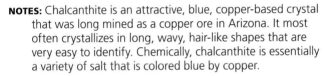

Occurrence

WHAT TO LOOK FOR: Long, hair-like crystals in shades of blue

SIZE: Natural chalcanthite is usually softball-sized and smaller

COLOR: Deep to light blue

OCCURRENCE: Uncommon

NOTES: Chalcanthite is an attractive, blue, copper-based crystal that was long mined as a copper ore in Arizona. It most often crystallizes in long, wavy, hair-like shapes that are very easy to identify. Chemically, chalcanthite is essentially a variety of salt that is colored blue by copper.

Care must be taken when adding chalcanthite to your collection. Since it is a salt, it is very easily affected by moisture—even humidity in the air—and will dissolve in water very quickly, which produces a blue, poisonous solution. In addition, natural chalcanthite contains iron, which makes the mineral very brittle. All of this makes chalcanthite very short-lived in collections. In fact, chalcanthite is so soluble that in Arizona's old copper mines, rainwater seeping through the earth and into mine shafts often dissolves the crystals, then precipitates new crystals on the mine's walls as the water dries up.

A prudent collector will take note that since chalcanthite solutions are very easily made, many of the commercially available crystals are man-made and should be regarded as fake, despite actually being more stable than natural chalcanthite. These man-made crystals are often large with good crystal faces and are grown on top of a solid rock matrix—all of which are not traits of natural chalcanthite.

WHERE TO LOOK: Mine dumps in the Patagonia Mountains, north-east of Nogales, and near Globe are good places to start. **67**

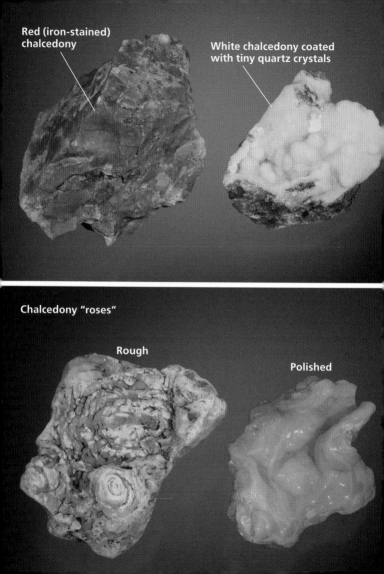

Red (iron-stained) chalcedony

White chalcedony coated with tiny quartz crystals

Chalcedony "roses"

Rough

Polished

Chalcedony

HARDNESS: 7 **STREAK:** White

Occurrence

ENVIRONMENT: Rivers, washes, mountains, buttes, roadcuts

WHAT TO LOOK FOR: Massive chunks (compact mineral concentrations) or coatings of hard, translucent material

SIZE: Chalcedony occurs massively and can be found in any size

COLOR: White to gray, blue, red to brown, pink, yellow

OCCURRENCE: Common

NOTES: Like chert and jasper, chalcedony is a variety of quartz with crystals that are so small that they cannot be seen without a microscope (these are known as microcrystals). However, chalcedony differs from chert and jasper in terms of its structure. Chert and jasper are opaque because their microcrystals are tightly packed together, while chalcedony's are arranged into parallel fibers, making them translucent. Chalcedony usually occurs massively and does not form visible crystals, though some specimens will be found with small crystal points of clear quartz (and some are coated with them). In addition, chalcedony sometimes forms "chalcedony roses;" these are "puddles" of chalcedony that formed when the mineral was still a liquid solution.

Pure chalcedony is white, but it occurs in many colors. Reds and browns are the most common colorations and are caused by iron staining the mineral. Because chalcedony, jasper and chert are varieties of quartz, they share its hardness, waxy surface luster and a tendency to fracture conchoidally (when struck, circular cracks appear). Translucency will distinguish chalcedony from chert or jasper.

WHERE TO LOOK: Chalcedony is very widespread, but a good place to look is anywhere there is exposed rock, especially basalt in the northeast corner of the state.

Chalcocite (gray)

Pyrite (gold-colored)

Quartz

Chalcocite (gray)

Chalcocite

HARDNESS: 2–2.5 **STREAK:** Shiny black

ENVIRONMENT: Mine dumps

WHAT TO LOOK FOR: Soft, dark gray metallic masses, often occurring with pyrite

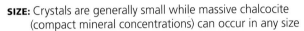
Occurrence

SIZE: Crystals are generally small while massive chalcocite (compact mineral concentrations) can occur in any size

COLOR: Dark lead gray, often tarnished to black

OCCURRENCE: Common

NOTES: Chalcocite is one of the most abundant copper minerals in the world and is the primary ore extracted in many of Arizona's copper mines. The dark gray mineral occurs most commonly as massive pieces, but it can also be found in compact, granular masses. Crystals, on the other hand, are much harder to come by. They are generally small and prismatic with striated (grooved) faces.

Chalcocite normally isn't confused with any other minerals. It is quite brittle, which is uncommon for a metallic material, and it is very abundant in copper-rich areas. These traits, combined with its dark gray color, are normally enough to identify it. However, tarnished bornite or chalcopyrite may appear very similar to chalcocite. Standard streak and hardness tests may help, because each mineral has a slightly different hardness and streak color.

Chalcocite is often found with other copper minerals, such as chalcopyrite, bornite and copper itself, but it is also commonly found with galena, pyrite and quartz.

WHERE TO LOOK: The mine dumps near Mammoth, as well as in the Sierrita Mountains, northwest of Nogales.

Blue tarnish

Chalcopyrite

HARDNESS: 3.5–4 **STREAK:** Greenish black

ENVIRONMENT: Mine dumps

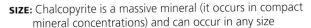

Occurrence

WHAT TO LOOK FOR: Metallic mineral with a colorful surface coating

SIZE: Chalcopyrite is a massive mineral (it occurs in compact mineral concentrations) and can occur in any size

COLOR: Bright to dark golden yellow, but orange, blue, purple, green or pink when tarnished; iridescent

OCCURRENCE: Common

NOTES: One of the most abundant copper minerals, especially in Arizona, chalcopyrite is an attractive, collectible mineral. Its bright golden yellow color is very similar to pyrite's, but chalcopyrite is much softer and is often found tarnished to iridescent shades of blue, green or purple. Pyrite is also commonly found as cubic crystals whereas chalcopyrite is generally in massive chunks. Another mineral that is very easily confused with chalcopyrite is bornite, which also tarnishes to similar iridescent colors. However, bornite is softer and tarnishes faster when freshly broken.

Chalcopyrite's tarnish is generally just a surface feature, and a fresh scratch or break will reveal a gold-colored interior. Fresh chalcopyrite could be confused with gold, but gold is extremely rare and less brittle than chalcopyrite.

Chalcopyrite is found with bornite, chalcocite, pyrite, sphalerite and galena.

WHERE TO LOOK: Mine dumps near Bisbee and south of Huachuca City, as well as in the Pinal Mountains, east of Phoenix and near Globe.

Hard, waxy surface

Arrowhead artifacts made from chert

Rough specimen

Chert

HARDNESS: 7 **STREAK:** White

Occurrence

ENVIRONMENT: Desert, buttes, roadcuts, rivers, washes

WHAT TO LOOK FOR: Hard massive chunks (compact mineral concentrations) that break conchoidally (circularly) and have a waxy appearance

SIZE: Chert is a massive mineral and occurs in any size

COLOR: White to gray, black

OCCURRENCE: Common

NOTES: Chert is a dense, hard, sedimentary rock that consists of microcrystalline quartz, much like jasper or chalcedony. In fact, chert and jasper are essentially the same thing—compact masses of tiny quartz crystals. Chalcedony differs slightly in that its microcrystals are arranged in a fibrous orientation and it is more translucent than chert or jasper. Chert is found in shades of gray, though sometimes bright white or dark black, whereas jasper tends to occur primarily in reds, browns and yellows. Since chert is a variety of quartz, it is easy to identify using the tests for identifying quartz. Like quartz, chert is very hard and it has a conchoidal fracture (when struck, circular cracks appear). It also is opaque and has a waxy surface luster. Of course, all these traits also apply to jasper, so the determining factor between those two is color. Black, fine-quality chert is known as flint, though few places in the United States actually produce chert of high enough quality to be considered flint. Chert has been used in Arizona for thousands of years by ancient peoples as tools, often chipped and broken into sharp points. While chert relics are beautiful, it is illegal to remove any from sites in Arizona without permission.

WHERE TO LOOK: Chert is found anywhere there is sedimentary rock.

Basalt

Chlorite

Cloudy greenish chlorite inclusions

Quartz crystal

Chlorite Group

HARDNESS: 2–2.5 **STREAK:** Colorless

ENVIRONMENT: Buttes, mountains, mine dumps, roadcuts

Occurrence

WHAT TO LOOK FOR: Soft green minerals, often lining vesicles (gas bubbles) in host rock

SIZE: Chlorite generally occurs in specimens smaller than your thumbnail

COLOR: Light to dark green

OCCURRENCE: Common

NOTES: The chlorite group encompasses a variety of soft green minerals that form in thin, scaly, often six-sided flakes. It is unnecessary for the non-geologist to differentiate between specific varieties, as the chlorites are very similar. Well-formed, bladed crystals are rare and chlorite normally occurs within other rocks and minerals in a variety of habits (forms). Most commonly, it appears as a greenish area within rocks such as basalt. This occurs because chlorite easily penetrates rock, replacing the minerals within. Chlorite is also notorious for turning quartz crystals cloudy, coloring them gray or green. Chlorite commonly lines vesicles (gas bubbles) within basalt; other vesicular minerals like chalcedony, calcite and zeolites sometimes have an outer coating of soft chlorite. Saponite, a soft clay material, can also have a similar green color and frequently lines vesicles, though confusing it with chlorite is unlikely. Saponite is much softer, doesn't crystallize the same way and is generally brown in color. Chlorite could also be confused with mica, especially in schists (where chlorite sometimes forms), but micas have a higher luster and more perfectly layered habit (form).

WHERE TO LOOK: Look anywhere there is ample basalt, namely the northern half of the state and around Flagstaff.

Rough chrysocolla

Transparent "gem chrysocolla"

Chrysocolla (blue)

Malachite (green)

Polished cabochons

Tenorite (black)

Cabochons courtesy of Mitchell Dale

Chrysocolla

HARDNESS: 2–4 **STREAK:** White to pale blue

Occurrence

ENVIRONMENT: Mine dumps, desert, mountains

WHAT TO LOOK FOR: Soft, blue-to-green masses filling cracks or cavities in other rock

SIZE: Chrysocolla can occur in any size, but is generally smaller than golf ball-sized specimens

COLOR: Blue to greenish blue

OCCURRENCE: Common

NOTES: Chrysocolla is one of Arizona's most colorful minerals and as such it is highly collectible. Its green and blue hues are not uncommon for copper-based minerals, but chrysocolla's color is frequently more intense than other related minerals. Its structure is microcrystalline, so you won't find any chrysocolla crystals. Instead, chrysocolla is often found in massive specimens (compact mineral concentrations) or botryoidal (grape-like) specimens, often filling vesicles or veins in rock. Specimens of chrysocolla kept in your collection for a long time will begin to desiccate, or dry out, which results in a brittle, and often cracked, specimen. Chrysocolla specimens should never be washed or wetted because the addition of water can result in the specimen crumbling and being destroyed completely. Turquoise is often confused with chrysocolla, but distinguishing the two is easy. Turquoise is a 6 in hardness, whereas chrysocolla is much softer. Chrysocolla also has a unique identifying trait—because it is a "dried-out" mineral, specimens will stick to your tongue. "Gem chrysocolla" is much harder and tests at a 7 on the Mohs hardness scale. This is actually quartz that has been colored green or blue by inclusions of chrysocolla.

WHERE TO LOOK: Mine dumps in the Bradshaw Mountains, south of Prescott, and the Santa Rita Mountains, southeast of Tucson.

Cinnabar vein

Quartz matrix

⚠ **Cinnabar**

HARDNESS: 2–2.5 **STREAK:** Brownish red

ENVIRONMENT: Mine dumps, mountains

WHAT TO LOOK FOR: Bright red veins or streaks in quartz or rock

Occurrence

SIZE: Most cinnabar veins are smaller than your thumbnail

COLOR: Light to dark red, pink

OCCURRENCE: Very rare

NOTES: Cinnabar contains mercury and is the only commercial source of the element. Since mercury, sometimes called "quicksilver," is well known to be detrimental to human health, great care should be taken when collecting cinnabar, despite the fact that cinnabar's sulfur content somewhat neutralizes the mercury. Cinnabar is very rare in Arizona and mostly occurs as veins within quartz, where its bright red color is easily recognized. Cinnabar crystals are rare worldwide and appear as translucent red points.

Most cinnabar mining has been halted due to adverse environmental effects. Therefore, finding cinnabar somewhere other than a mineral museum or store is extremely difficult or impossible without permission to investigate mines where it was dumped.

Cinnabar veins, especially when surrounded by a large mass of another mineral or rock, are relatively safe to handle. However, it should be noted that some cinnabar veins have been known to contain beads of pure mercury. In addition to quartz, these veins can occur with chalcedony, cristobalite (opal) and barite. Always remember to wash your hands thoroughly after handling cinnabar and avoid inhaling any dust produced when working with it.

WHERE TO LOOK: The Dome Rock Mountains, west of Quartzsite.

"Moqui marbles"

Concretions

HARDNESS: N/A **STREAK:** N/A

ENVIRONMENT: Desert, buttes, washes

WHAT TO LOOK FOR: Unusually round, hard balls of rock

Occurrence

SIZE: Concretions can be pea-sized to softball-sized and rarely larger

COLOR: Brown to black, tan and red

OCCURRENCE: Uncommon

NOTES: A concretion is an unusually round body of rock that formed when sediment collected around a central core. Sometimes, the core is organic material, such as part of a plant, while other times it is a compact ball of sandstone. There are many different kinds of concretions around the world, and while "Moqui marbles" are normally considered to be from Utah, there is at least one locality in northern Arizona where they can be found.

"Moqui marbles," as they are known locally, are concretions made of iron. They formed when solutions containing iron oxides seeped through sandstone or mudstone and hardened, creating round, compact, grainy-textured masses. These concretions often have a core of lighter-colored sand inside. They are not particularly valuable unless uniquely shaped.

Concretions, specifically the "Moqui marbles," are easy to identify. Be on the lookout for dark, hard, round masses that are generally golf ball-sized. It's possible to confuse a concretion with a geode, another round mineral formation, but geodes often contain crystals or agate and concretions will not.

WHERE TO LOOK: Sandy areas in the far north of Arizona.

Conglomerate

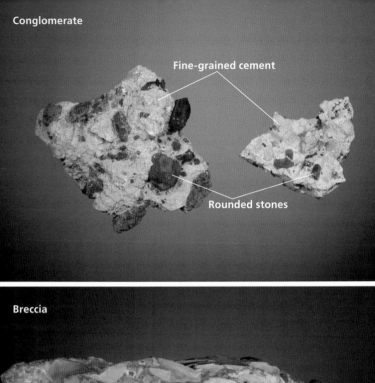

Fine-grained cement

Rounded stones

Breccia

Broken, angular fragments

Conglomerate/Breccia

HARDNESS: N/A **STREAK:** N/A

ENVIRONMENT: Desert, buttes, mountains, roadcuts

Occurrence

WHAT TO LOOK FOR: Rock that appears to be made of many smaller rocks or chunks of rocks

SIZE: Both conglomerate and breccia can be found in any size

COLOR: Varies greatly

OCCURRENCE: Common

NOTES: Both conglomerate and breccia are unique rock types in that they contain many smaller bodies of rock, rather than mineral crystals. The rocks contained within masses of conglomerate or breccia can really be any other rock type, depending on how the rock formed. Because of their great variability, hardness and streak tests do not apply.

Conglomerates are composed of a fine-grained rock or sand acting as a cement that hardens around rounded, unbroken, individual stones. This often occurs underwater or in other sedimentary environments. When weathered, the stones contained within it often stand up off the surface of the rest of the rock because they are harder. Most conglomerates are not very interesting or valuable, but large specimens of the material can be used as a decorative construction stone.

Breccia is formed of sharp, angular fragments of broken rock that are cemented together. This is often a result of volcanic activity, earthquakes or any other violent action that can crush rock. The broken pieces are then solidified by a fine rock or mineral that works its way between the fragments and hardens. Tuff, a variety of hardened volcanic ash, is often the cement in volcanic breccias.

WHERE TO LOOK: Anywhere there is exposed rock.

Botryoidal (grape-like) conichalcite (green)

Conichalcite

HARDNESS: 4.5 **STREAK:** Green

ENVIRONMENT: Mine dumps

Occurrence

WHAT TO LOOK FOR: Tiny, bright green crystals on or in another mineral or rock

SIZE: Most crystals are tiny and just millimeters in width

COLOR: Bright grass green to dark olive green

OCCURRENCE: Uncommon

NOTES: Conichalcite is an attractive, hard-to-find collectible in Arizona. Its crystals are small and fibrous, but you'll most likely encounter this mineral as little botryoidal (grape-like) crusts on the surface of another mineral or rock. Each individual sphere of conichalcite is generally very small—no more than just a millimeter or two in diameter. Their small size can make them difficult to find, especially when a small number of conichalcite crystals are growing alongside other green minerals, like malachite. As with many of Arizona's green minerals, conichalcite gets its color from its copper content. Despite being one of many similarly colored minerals, however, conichalcite's shades of green are often much more vivid. In fact, its color is one of its best identifying features.

Conichalcite occurs with malachite and azurite, which are other copper-based minerals, though telling them all apart should be easy. It would be more likely that you would confuse conichalcite with brochantite or atacamite, two copper minerals rarer than malachite and azurite. Conichalcite's crystal habit (form) and color are normally enough to tell them apart, but when its color happens to be on the darker side you'll have to rely on crystal structure and hardness tests.

WHERE TO LOOK: Mine dumps near the cities of Bagdad, Bisbee and Globe.

Copper nugget

Copper crystal

Copper in selenite

Copper crystal in calcite (white)

Copper

HARDNESS: 2.5–3   **STREAK:** Metallic red

ENVIRONMENT: Mine dumps, mountains, rivers

Occurrence

WHAT TO LOOK FOR: Reddish metal, often green or black with tarnish

SIZE: Most copper is found as small nuggets or crystals no larger than a golf ball, but many specimens can be much larger

COLOR: Bright metallic red when fresh, dark brown to green or black when oxidized

OCCURRENCE: Uncommon

NOTES: Copper is perhaps Arizona's most important mineral geologically, if not commercially. It has been mined all over the state for a century and is responsible for the many colorful, collectible minerals that derive from weathered copper deposits. Chalcopyrite, malachite, azurite, cuprite, aurichalcite, brochantite and a suite of other minerals are all formed of copper. Of course, copper itself is highly collectible, too.

Copper's most common crystal shape is a cube, though you're most likely to find copper crystals as elegant fern-like branches of individual pointed crystals. These branching shapes actually consist of many smaller, distorted crystals. However, like most minerals, crystals are much rarer than massive varieties. Copper nuggets, ores and sheets are still available in the dumps left over from old mines, though after decades of collectors digging through the piles, specimens are scarce. (In addition, most mines are private property.) You're very unlikely to confuse copper with any other mineral as its metallic red color is distinctive. Tarnished copper is green to blue or black, but a fresh scratch should reveal its true color.

WHERE TO LOOK: Mine dumps near Globe, Clifton and Tombstone.

White opal

Pink opal

Blue opal

Rare "fire" in a specimen of blue opal

Cristobalite (Opal)

HARDNESS: 5.5–6.5 **STREAK:** White

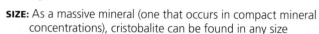
Occurrence

ENVIRONMENT: Deserts, buttes, roadcuts

WHAT TO LOOK FOR: Glass-like masses with no crystal structure

SIZE: As a massive mineral (one that occurs in compact mineral concentrations), cristobalite can be found in any size

COLOR: White or gray are most common but also blue or pink

OCCURRENCE: Rare

NOTES: Cristobalite, more commonly known as opal, is unique in that it does not form crystals, but rather forms in massive chunks that fill cavities and veins within rock. It can also form botryoidally (grape-like) or stalactitic (cone-shaped), but these types are less common. Opal has a nearly identical composition to quartz, but opal contains more water, which makes it softer. Since quartz is one of the few minerals that could be confused with opal, a hardness test is a good indicator.

Rarely, opal will exhibit an iridescence, or color play, and such a specimen is called "precious opal." This happens because of microscopic, uniformly-sized spheres of silica (quartz material) aligned perfectly with each other, which then reflect light within themselves. But all opal, precious or not, is often short-lived within collections. Because of the excess water within opal, it easily dries out on exposure to air. This results in the surface of the specimen becoming covered in tiny cracks, which cause the piece to crumble over time. Serious opal collectors place their specimens in jars filled with mineral oil in order to help them retain moisture for as long as possible.

WHERE TO LOOK: The area around Sunset Crater, north of Flagstaff, and the Muggins Mountains, east of Yuma, are known sources of opal.

Copper

Cuprite (red)

Limonite (brown)

Polished cabochon with copper

Chalcotrichite (red) on chrysocolla (green)

Cuprite

HARDNESS: 3.5–4 **STREAK:** Brown red

Occurrence

ENVIRONMENT: Mine dumps

WHAT TO LOOK FOR: Metallic reddish crystals, often forming with copper

SIZE: Individual cuprite crystals are generally smaller than a pea

COLOR: Red to reddish black

OCCURRENCE: Uncommon

NOTES: Cuprite is an uncommon but attractive copper-based mineral that forms beautiful, deep red translucent crystals, often atop copper itself. When finding cuprite, you're most likely going to see little red square-sided crystals that have a bright, metallic luster. Occasionally, cloudy, grayish cuprite crystals will appear to be opaque because of their high reflectivity, and to the untrained eye, these cuprite specimens will look like a red, metallic material.

Cuprite's color and hardness are normally enough to distinguish it from other minerals. The variety of cuprite known as chalcotrichite is also fairly easy to identify. Chalcotrichite crystals are tiny, needle-like prisms that form in aggregates (groups) on a matrix. These delicate aggregates are often so tightly formed that the individual crystals reflect light in such a way that the specimen as a whole greatly resembles velvet. Chalcotrichite's bright red color and crystal habit are easy to recognize.

Cuprite often occurs with copper and other copper-based minerals, such as azurite, malachite and tenorite, but it can also occur with calcite and quartz.

WHERE TO LOOK: The best places to start are the mine dumps around Bisbee or Globe, or the Dripping Spring Mountains, south of Globe.

Limestone

Tree-like surface formations

Dendrites

HARDNESS: N/A **STREAK:** N/A

ENVIRONMENT: Desert, buttes, mountains

WHAT TO LOOK FOR: Dark, tree-like mineral growths on the surface of rocks

Occurrence

SIZE: Dendrites are usually no more than an inch or two long

COLOR: Dark gray

OCCURRENCE: Uncommon

NOTES: Deriving from the Greek word for "tree," the term dendrite refers to a specific type of mineral growth, rather than an individual mineral itself. As such, dendrites appear as tree-like coatings on the surface of rock, primarily limestone. These "branches" formed when various oxides of the element manganese seeped into the tiny cracks and spaces in a body of rock and dripped downward.

Dendrites are essentially only two-dimensional. They are a very thin coating on the surface of the rock and do not go inward into the stone. Therefore, it is easy to destroy a dendrite simply by trying to wash or clean it, polishing it or even just rubbing it too hard with your hands. However, dendrites can occur on different types of rock other than limestone, and even within other minerals, such as chalcedony. Manganese dendrites within chalcedony are often referred to as "moss agates," and in this case the dendrites are protected within the very hard chalcedony.

Dendrites are often marketed as fossils, but this is incorrect. They may look like a fossilized plant, but they're inorganic.

WHERE TO LOOK: The hills around Tombstone and Tucson, as well as any place where there is exposed limestone.

Fine-grained gabbro-like appearance

White "fuzzy" feldspar bunches

Close-up of "fuzzy" feldspar bunches in diabase

Diabase

HARDNESS: >5.5 **STREAK:** N/A

ENVIRONMENT: Mountains, buttes, roadcuts, rivers

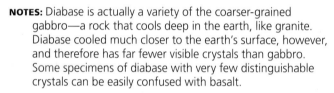

Occurrence

WHAT TO LOOK FOR: Dark, gray rock containing few small crystals

SIZE: Diabase can occur in any size

COLOR: Gray or black with differently colored spots

OCCURRENCE: Common

NOTES: Diabase is actually a variety of the coarser-grained gabbro—a rock that cools deep in the earth, like granite. Diabase cooled much closer to the earth's surface, however, and therefore has far fewer visible crystals than gabbro. Some specimens of diabase with very few distinguishable crystals can be easily confused with basalt.

Many times, the crystals within diabase appear "fuzzy" or "out-of-focus." This is because the feldspar began to crystallize before the rest of the rock cooled. The other, darker minerals then filled in around the crystals and prevented them from growing more. This is the opposite of what usually happens in most other dark volcanic rocks. Generally, it is the darker minerals, not the feldspar, that form first, as is the case with basalt.

Both diabase and basalt contain many of the same minerals as gabbro, hence their similar appearance. The best way to distinguish between the two is to look for diabase's light, sometimes poorly-formed crystals surrounded by dense, dark rock.

WHERE TO LOOK: Most diabase can be found in areas of volcanic rock, such as the mountains around Flagstaff and Clifton.

Diopside

HARDNESS: 5–6 **STREAK:** Greenish gray

Occurrence

ENVIRONMENT: Mountains, buttes, roadcuts, mine dumps

WHAT TO LOOK FOR: Green crystals embedded within lighter-colored rock, such as marble

SIZE: Most crystals are pencil-width and no longer than two inches

COLOR: Light to dark green

OCCURRENCE: Common

NOTES: Like augite, diopside is a member of the pyroxene mineral group. Pyroxenes are dark minerals that often form within rocks, such as granite. Diopside is no exception, and its crystals are most frequently found embedded in other rock, especially marble. Fine diopside crystals are short, prismatic, and are of a rich, translucent green color. However, the likelihood of finding such a well-formed crystal is low. Most diopside appears as long green "smudges" on marble, sometimes very vaguely crystal-shaped or not at all.

Epidote and olivine are similarly colored and can be confused with diopside, but it's easy to tell them apart. Epidote generally has a yellow-green color and often grows more freely than diopside; in addition, epidote's crystals are structurally different than diopside's. Olivine is harder and does not occur in marble. Though marble is the primary environment for diopside, it can also occur in metamorphosed (altered) limestones.

WHERE TO LOOK: The mine dumps south of Globe and east of Phoenix, as well as the Santa Rita Mountains and the Sierrita Mountains, both south of Tucson, are good places to look.

Chrysocolla

Quartz (white)

Dioptase

Dioptase

Dioptase (green) on quartz (white)

Dioptase

HARDNESS: 5 **STREAK:** Pale green blue

ENVIRONMENT: Mine dumps

WHAT TO LOOK FOR: Small green crystals coating other rocks and minerals

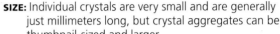

Occurrence

SIZE: Individual crystals are very small and are generally just millimeters long, but crystal aggregates can be thumbnail-sized and larger

COLOR: Emerald green to teal

OCCURRENCE: Uncommon

NOTES: Dioptase is one of Arizona's many copper-related minerals, but it is also one of the most unique. While minerals like azurite, malachite, chrysocolla and chalcanthite are all colored blue or green by copper, none have quite the same distinct translucent teal color as dioptase. For many collectors, this color alone is enough to identify the mineral. Its crystals are generally very small and form clusters on the surface of another mineral, such as quartz, limonite or chrysocolla; in Arizona, pea-sized crystals of dioptase are considered large.

Small malachite crystals can possibly be confused with dioptase if the colors are similar. In that case, a hardness test, while difficult to conduct with such small crystals, will help differentiate the materials, as dioptase is harder. In fact, dioptase is harder than most of the similarly colored copper minerals.

Dioptase can be found in rock cavities in copper-rich regions.

WHERE TO LOOK: Mine dumps in the Mammoth area, Superior area and the Clifton area, all east of Phoenix, have yielded dioptase crystals.

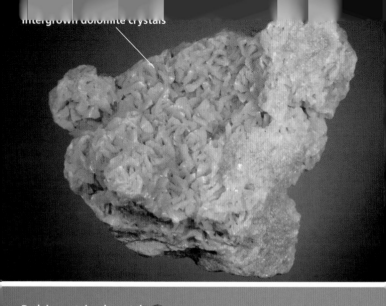

Intergrown dolomite crystals

Red, iron-stained crystals

Dolomite

HARDNESS: 3.5–4 **STREAK:** White

ENVIRONMENT: Mine dumps, buttes, mountains, roadcuts

Occurrence

WHAT TO LOOK FOR: Light-colored crystals, often with curved faces and a high luster

SIZE: Individual crystals are thumbnail-sized and smaller

COLOR: Colorless to white, also yellow, pink, red or brown, depending on impurities

OCCURRENCE: Common

NOTES: Dolomite is a common, light-colored mineral that most often forms in small crystals with curved, or "saddle-shaped," faces. Crystals are most commonly flesh-colored or white, but can occur in any number of other colors, depending on impurities that may be within the mineral. Brown or red are common color variations that result from the inclusion of iron.

While crystals are the most desirable specimens for collectors, many minerals occur in a variety of forms. Dolomite is no exception. Since dolomite frequently grows within cavities in limestone deposits, it can sometimes alter the limestone itself. The resulting dolomite-rich limestone is called dolomite rock, or sometimes "dolostone," which is often used as a decorative material. Dolomite rock differs little from limestone, but is sometimes pink colored. Dolomite can resemble calcite, but dolomite's curved crystal faces are normally enough to differentiate it. If crystals are not present, a hardness test should be used, since dolomite is harder than calcite.

WHERE TO LOOK: The Sierrita Mountains, south of Tucson, and the Tombstone Hills, west of Tombstone, are known dolomite locations.

Rough specimen

Polished specimen

Polished cabochons

Dumortierite

HARDNESS: 7 **STREAK:** White

ENVIRONMENT: Mountains, roadcuts, mine dumps

Occurrence

WHAT TO LOOK FOR: Deep blue, hard, massive chunks

SIZE: Dumortierite crystals are no more than an inch or two long, but massive dumortierite can occur in any size

COLOR: Dark blue to purple or black

OCCURRENCE: Rare

NOTES: There are very few locations in Arizona where you can find dumortierite, which is a hard, dark blue mineral widely used in jewelry. While it does occur elsewhere as delicate fibrous crystals, Arizona's dumortierite is found as massive chunks (compact mineral concentrations) that are so compact that virtually no crystal structure can be seen. It often occurs alongside the equally hard quartzite, sometimes intermixed together.

Because dumortierite is quite hard, it polishes well and can make for attractive specimens. It occurs in shades of blue or purple and the brighter and more vivid the color, the more valuable and collectible it becomes. Few other minerals in Arizona are colored quite the same way, making the coloration of the mineral a distinguishing characteristic.

Most of Arizona's dumortierite is found near the California border in the same rock formation that produces kyanite and andalusite. While kyanite also exhibits bluish hues, it is not as hard nor as opaque as the dumortierite found there. Also, as mentioned above, quartzite is often found with dumortierite and has a similar hardness, though quartzite is normally much lighter in color and grainier in appearance.

WHERE TO LOOK: The Trigo Mountains, north of Yuma, as well as the area around Lake Havasu are good places to look.

Epidote

HARDNESS: 6–7 **STREAK:** Colorless to gray

Occurrence

ENVIRONMENT: Mountains, buttes, roadcuts, mine dumps

WHAT TO LOOK FOR: Flat, elongated, yellow-green crystals with striated (grooved) sides

SIZE: Crystals can be quite large but are generally thumbnail-sized and smaller

COLOR: Yellow-green to greenish black

OCCURRENCE: Common

NOTES: Epidote is a hard, dark-colored mineral with a yellow-green color that is quite distinctive and it is not easily confused with other minerals. Its crystals are often flat and long and have striated (grooved) faces; even poorly formed or broken crystals exhibit this trait. The exception is when epidote fills a vesicle (gas bubble) within another rock and its crystal structure is not easily visible. However, even if the crystal structure is difficult to see, the yellow-green color will probably be enough to identify it.

Epidote occurs in a number of environments, but granite pegmatites yield the largest and most well-formed crystals. A pegmatite is the lowest, and slowest-cooling, portion of a granite formation. This allows the various minerals in the magma to crystallize, sometimes in very large specimens. Some of the world's largest crystal specimens form in pegmatites. Epidote can also as occur as crystals in limestone regions, but the most common form will be small crystals and amygdules (material formed within a gas bubble in rock) within cavities in rock. Epidote occurs with zeolites, calcite and chlorite.

WHERE TO LOOK: The Bradshaw Mountains, south of Prescott, and the mine dumps around Quartzsite are known locations for good crystals.

Orthoclase feldspar crystal

Blunt, rounded edges

Feldspars in granite

Orthoclase feldspar

Microcline feldspar

Orthoclase feldspar

Feldspar Group

HARDNESS: 6–6.5 **STREAK:** White

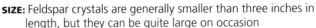

ENVIRONMENT: Found in all environments

WHAT TO LOOK FOR: Generally light-colored, opaque minerals in granites or as crystals

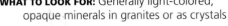
Occurrence

SIZE: Feldspar crystals are generally smaller than three inches in length, but they can be quite large on occasion

COLOR: White to gray, sometimes pink, yellow, or dark gray

OCCURRENCE: Very common

NOTES: The term "feldspar" refers to a group of closely related minerals that contain aluminum mixed with either potassium, calcium or sodium. The feldspar group as a whole is considered the most common mineral on earth, making up approximately sixty percent of the earth's crust. This extremely important group of minerals is found in many different forms. Most commonly, you'll find feldspars as a part of rocks, especially granites, in which the feldspars are generally white or pink. Feldspars found this way often exhibit striations (grooves) on their faces and can reflect light differently than the rest of the granite.

Feldspar crystals are much less common than feldspars found within rocks. While each variety of feldspar can crystallize differently, most are found as broad, blocky, opaque crystals that are often poorly formed. All feldspars are generally white to gray, but can be pink or flesh-colored as well as brown, depending on impurities.

The abundance of feldspar, its lighter colors and its hardness make feldspar easy to identify. The most abundant feldspar types are orthoclase, microcline and plagioclase.

WHERE TO LOOK: The Dome Rock Mountains, 5 miles west of Quartzsite, have produced good crystals, but feldspars can be found everywhere.

Rough fluorite

Color banding

Fluorite octahedron

Quartz (white)

Fluorite (purple)

Fluorite

HARDNESS: 4 **STREAK:** White

ENVIRONMENT: Mine dumps, mountains, roadcuts

Occurrence

WHAT TO LOOK FOR: Glassy masses or crystals with cubic or octahedral (eight-sided) formations

SIZE: Fluorite can occur in virtually any size, but crystals are generally smaller than a golf ball

COLOR: Colorless to white or yellow, green, purple, blue, or black

OCCURRENCE: Common

NOTES: Fluorite is a colorful, collectible mineral found in many places around Arizona. It is often white or colorless, but bright purples and greens are not uncommon and make fine collector's specimens. Many times, a sample of fluorite has several colors arranged into bands. Fluorite crystals are most commonly cubic (box-like), but some are more famously octahedral (eight-faced). Fluorite has a perfect octahedral crystal structure; in fact, any piece of fluorite will yield a perfect octahedron when carefully broken. This action is called cleavage, or the tendency of a mineral to break along its crystalline molecular structure.

Fluorite can be confused with quartz or calcite, but distinguishing them is easy. Quartz is much harder than fluorite while calcite is softer.

Fluorite is usually fluorescent under ultraviolet light. In fact, the word "fluorescent" is derived from fluorite itself, as fluorite was the first mineral identified that glowed under ultraviolet light.

WHERE TO LOOK: Mine dumps around Globe and Superior, as well as the Little Dragoon Mountains, north of Tombstone.

Petrified wood

Wood-grain texture

Polished petrified wood

Specimen courtesy of Pat McMahan

Various fossilized corals in limestone

Specimens courtesy of Mitchell Dale

Fossils Group

HARDNESS: N/A **STREAK:** N/A

ENVIRONMENT: Desert, buttes, roadcuts

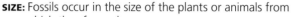

Occurrence

WHAT TO LOOK FOR: Rocks or minerals that resemble a once-living plant or animal

SIZE: Fossils occur in the size of the plants or animals from which they formed

COLOR: Varies greatly

OCCURRENCE: Rare

NOTES: Fossils formed when the bodies of ancient animals and plants slowly turned to rock. Minerals seeped into the cells of the once-living organism and replaced the organic tissue with rock. Fossils are quite easy to identify once discovered. Be on the lookout for strange impressions and odd shapes that resemble living things. Arizona has many fossils, though none are more notable than Arizona's famous petrified wood. The Petrified Forest, near Holbrook, contains thousands of fossilized trees, now found as rock "logs," many of which still exhibit the ancient trees' growth rings. Others were replaced by jaspers, which produced beautiful, brightly colored fossilized wood. The area is a protected park, however, and removing even the smallest piece is illegal.

Limestone, a rock that Arizona has plenty of, is a prime environment for fossils. Within it, you'll find a wide range of specimens, but primarily traces of marine life forms, such as the shells of clams or branches of coral. Other, more exciting, fossils are also found in Arizona, including the remains of mastodons. Such fossils are much rarer.

WHERE TO LOOK: Roadcuts along highways 64 and 260, as well as the area south of Clifton, are known for fossils.

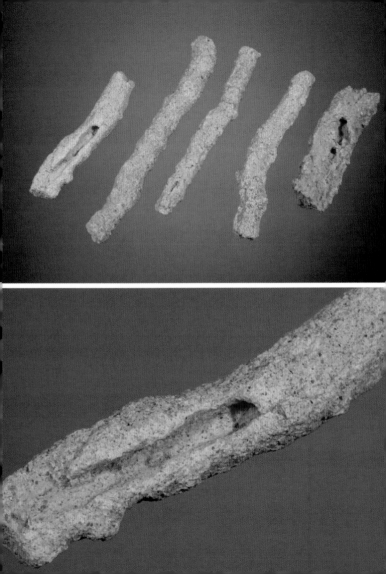

Fulgurite

HARDNESS: N/A **STREAK:** N/A

Occurrence

ENVIRONMENT: Desert, buttes

WHAT TO LOOK FOR: Hard branch-like tubes of sand

SIZE: Fulgurites are generally pencil-width and no longer than a few inches

COLOR: Varies, but most often tan or brown

OCCURRENCE: Rare

NOTES: Fulgurites are rare and unique collectibles, but they aren't actually a mineral. They form when lightning strikes a body of sand rich in quartz. The electricity from the bolt of lightning goes into the ground, melting the sand and turning it into glass. This produces long, hollow tubes that resemble tree branches. Fulgurite is often called "petrified lightning" by collectors and rock shops. While it may just be a clever nickname, it actually isn't such a bad description.

As mentioned above, fulgurite isn't actually a mineral. That's because fulgurites are made of lechatelierite, a type of silica (quartz) glass that has no crystal structure, and by definition, a mineral must have a defined, solid crystal structure. Fulgurites are hard to find because when the melted glass cools, the sand surrounding it sticks to it. Therefore the fulgurites are the same color as the sand it is found in. Keeping an eye out for hard, long masses of sand that resemble tree branches is the best way to find one. Also, using a rake to search loose sand is a good idea.

WHERE TO LOOK: Fulgurite formation depends entirely on where lightning strikes, so anywhere with large areas of sand is a good place to look.

Water-worn surface

Large, reflective, plagioclase feldspar crystal

Coarse, granite-like texture

Reflective crystals

Dark, mottled appearance

Gabbro

HARDNESS: 5.5 **STREAK:** N/A

ENVIRONMENT: Mountains, roadcuts, rivers, buttes

Occurrence

WHAT TO LOOK FOR: Dark greenish gray rock containing many glassy crystal fragments

SIZE: Gabbro can be found in any size

COLOR: Dark gray to green with lighter colors mottled throughout

OCCURRENCE: Common

NOTES: Gabbro is a rock that forms in much the same way as granite but has a very different combination of minerals. Both rocks formed deep within the earth and cooled slowly, allowing the various minerals within them to form large, visible crystals. Gabbro is essentially the same as diabase but cooled more slowly and therefore has more visible crystals. The plagioclase feldspar crystals that make up gabbro are often so well formed that you can observe their crystal shape. The feldspar crystals are also highly reflective at certain angles when the rock is rotated. The other primary ingredient of gabbro is augite, a member of the pyroxene group, which are dark, rock-forming minerals.

While similar to granite in many ways, gabbro weathers much faster and can be found as crumbly pieces. It also breaks down often in rivers or washes, where it is found as rounded, worn stones.

Gabbro is generally easy to identify, as it is much darker than any granite, has much larger grains and crystals than diabase or basalt, and its greenish color is unique to rocks of its type in Arizona. In addition, the feldspars in gabbro can make freshly-broken specimens appear "sparkly," as each crystal reflects light individually.

WHERE TO LOOK: Gabbro will be found in areas of volcanic rock, such as the Flagstaff and Clifton areas.

Galena (metallic gray)

Cubic structure

Anglesite

⚠ Galena

HARDNESS: 2.5 **STREAK:** Dark gray

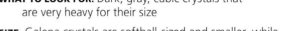

ENVIRONMENT: Mine dumps, mountains

WHAT TO LOOK FOR: Dark, gray, cubic crystals that are very heavy for their size

Occurrence

SIZE: Galena crystals are softball-sized and smaller, while massive galena (compact mineral concentrations) can occur in any size

COLOR: Dark lead-gray

OCCURRENCE: Common

NOTES: Galena is the world's primary lead ore; it is the most common lead-based mineral and occurs in many different mineral environments. It is also relatively easy to identify because of its color, luster and high specific gravity (it feels very heavy for its size). Crystals of galena are not uncommon and are mostly cubic (box-like), often occurring in elaborate formations of stacked or layered cubes. Since galena is so abundant, it also occurs massively as chunks that can be very large in size. Galena often has a very bright, metallic luster when freshly broken and is easy to recognize.

Dull, gray galena can greatly resemble chalcocite, and since the two minerals' hardness and streak are so similar, it can be very difficult to tell the two apart. Galena's specific gravity is higher, but its cubic structure is the best way to distinguish it. If no cubic crystals are present, break the specimen. Because of galena's cubic crystal structure, it will break at ninety degree angles (called cubic cleavage) whereas chalcocite does not. Galena often occurs with quartz, pyrite, chalcopyrite, anglesite and cerussite.

WHERE TO LOOK: The Copper Creek District, about 50 miles northeast of Tucson, as well as the Bradshaw Mountains, south of Prescott.

Andradite

Almandine

Andradite

Specimens courtesy of Mitchell Dale

Pyrope

⅛" (3 mm)

Grossular

Spessartine

Spessartine

Andradite

Garnet Group

HARDNESS: 6.5–7.5 **STREAK:** Colorless

Occurrence

ENVIRONMENT: Mountains, buttes, mine dumps, roadcuts

WHAT TO LOOK FOR: Hard, angular crystals embedded within other rock

SIZE: Garnets are generally smaller than an inch in size, but some can be golf ball-sized

COLOR: Brown to green, red to purple, black

OCCURRENCE: Uncommon

NOTES: Garnets are a group of very closely related minerals that can all generally be found as small, well-formed crystals. There are many varieties of garnets, but there are five prominent types most common in Arizona: Almandine (which occurs in brown or deep red), andradite (green or brown) and spessartine (deep, rich purple) are the three most abundant, but pyrope (bright red to brown) and grossular (green to dark brown) are also available. Most are a result of metamorphic activity causing heat and pressure within bodies of rock, but some garnet varieties form in volcanic rocks, while yet others form in granite pegmatites (very coarse granite formations).

Most garnet crystals occur as balls exhibiting many faces, not unlike a soccer ball. Others form blocky, angular masses. Garnet is very hard and often protrudes from the host rock, because the host rock is softer and weathers more easily. All are easy to identify, thanks to their characteristic color, hardness and small, often well-formed crystals.

WHERE TO LOOK: The best garnet locations are at Stanley Butte and Comb Ridge, but they are on reservations and you need permission to collect there.

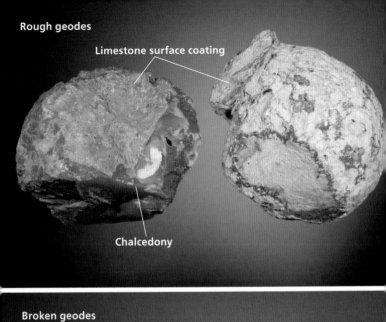

Rough geodes

Limestone surface coating

Chalcedony

Broken geodes

Quartz crystals

Limestone exterior

Geodes

HARDNESS: N/A **STREAK:** N/A

ENVIRONMENT: Desert, buttes, roadcuts

WHAT TO LOOK FOR: Round rock balls that are hollow inside when broken

SIZE: Most geodes are softball-sized and smaller

COLOR: White to gray or brown externally; colorless to white internally

OCCURRENCE: Uncommon

Occurrence

NOTES: Geodes are famous for their ability to surprise collectors. While on the outside they appear to be round balls of soft rock, inside they may contain beautiful crystals or agate. Many geodes are oblong or less than perfect spheres, yet some specimens are remarkably round. And while they can form in a wide variety of environments, geodes from Arizona generally form in cavities within limestone deposits or other sedimentary rocks. These cavities, much like some vesicles in basalt, begin to fill with hard quartz or chalcedony and crystallize inward. The majority of geodes contain coarse, white or colorless quartz crystals, whereas very fine geodes contain banded agate. Geodes are "camouflaged" because their exteriors resemble the rock they formed in, making them difficult to find. Since they can form in limestone beds, and since Arizona is very rich with limestone, keep a sharp eye out for pieces of limestone that just seem "too round." Once you have found a geode, however, it is easily identified because breaking it open will reveal the goodies inside. But be careful—very hollow geodes tend to crush, rather than crack open. On the other hand, geodes completely filled with quartz or chalcedony will be very difficult to break.

WHERE TO LOOK: The San Francisco River area near Clifton and the desert areas around Signal are known locations.

Layered minerals

Worn-down surface

Almandine garnets

Layered minerals

Gneiss

HARDNESS: N/A **STREAK:** N/A

ENVIRONMENT: Mountains, roadcuts, washes

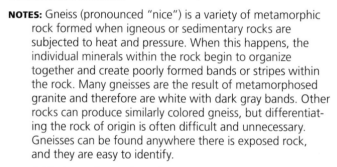

Occurrence

WHAT TO LOOK FOR: Coarse rock with poorly formed bands

SIZE: Gneiss can occur in any size

COLOR: Varies greatly, but primarily black, white, gray or brown

OCCURRENCE: Common

NOTES: Gneiss (pronounced "nice") is a variety of metamorphic rock formed when igneous or sedimentary rocks are subjected to heat and pressure. When this happens, the individual minerals within the rock begin to organize together and create poorly formed bands or stripes within the rock. Many gneisses are the result of metamorphosed granite and therefore are white with dark gray bands. Other rocks can produce similarly colored gneiss, but differentiating the rock of origin is often difficult and unnecessary. Gneisses can be found anywhere there is exposed rock, and they are easy to identify.

Schist is another variety of metamorphosed layered rock which can greatly resemble gneiss. In fact, they can often be the same thing, as the terms "gneiss" and "schist" are somewhat interchangeable. By definition, a rock is considered gneiss if less than half of its minerals are arranged into layers. This means that many of its minerals are still "mixed up" as they were when it was the original rock. Schist, on the other hand, has more than half of its minerals arranged into layers and exhibits less of the original rock's appearance. However, with such vague descriptions, the labelling of specimens is often left to the collector.

WHERE TO LOOK: Gneiss will be found in any metamorphic areas, which can include mountain ranges.

Massive goethite

Goethite coated in tiny quartz crystals

Botryoidal surface structure

Goethite

HARDNESS: 5–5.5 **STREAK:** Yellow to brown

ENVIRONMENT: Mine dumps, buttes, roadcuts

WHAT TO LOOK FOR: A massive brown metallic mineral with yellow oxidation

Occurrence

SIZE: As a mineral that generally forms massively (in compact mineral concentrations), it can be found in any size

COLOR: Yellow-brown to black

OCCURRENCE: Common

NOTES: Goethite is the world's second-most important iron ore, after hematite. In Arizona, goethite is extremely prevalent in nearly all well-mineralized (mineralogically diverse) regions, especially those rich in metallic minerals. Goethite crystals are rare and the mineral is most commonly found massively as botryoidal (grape-like) crusts colored in shades of brown or gray. It often occurs with quartz or calcite.

Goethite can easily be confused with both hematite and limonite. Hematite, another iron ore, is also often gray and botryoidal, but when hematite oxidizes, it turns to shades of red, whereas goethite turns yellow. If your specimen isn't oxidized at all, a streak test will easily tell the two minerals apart, as hematite's streak will be red and goethite's will be yellow. Limonite (massive iron oxide) is a yellow-brown color that looks like oxidized goethite. An easy way to tell the two apart is to look at structure; goethite exhibits a fibrous internal structure, whereas limonite has virtually no structure at all.

WHERE TO LOOK: The Sierrita Mountains and the Patagonia Mountains, both south of Tucson, and the Dripping Spring Mountains, east of Phoenix, all have mine dumps where goethite can be found.

Quartz

Gold flakes

Placer gold nugget

Rounded, bent surface

Specimen courtesy of Hallie Edwardson

Gold

HARDNESS: 2.5–3 **STREAK:** Metallic yellow

ENVIRONMENT: Rivers, mine dumps

WHAT TO LOOK FOR: Bright yellow metal generally found within quartz or in river beds

Occurrence

SIZE: Gold is normally found as very small, thin flakes

COLOR: Golden yellow

OCCURRENCE: Very rare

NOTES: Few minerals have excited people throughout history like gold. Its rarity, bright yellow color and resistance to corrosion make it a valuable addition to your mineral collection. Crystals are extremely rare because gold normally occurs as veins within other rocks or minerals, particularly quartz. These specimens are normally tiny flakes completely encased by the host rock. And since gold veins are generally mined commercially, collecting these specimens can be difficult.

Most collectors seeking gold turn to placer deposits, which are accumulations of dense, heavy, and often valuable minerals at the bottom of rivers or streams. Since gold has a very high specific gravity (it feels very heavy for its size), even very small pieces are trapped in holes and depressions underwater. While still very rare, weathered nuggets of gold are more commonly found this way.

Gold can be confused with pyrite, though pyrite, often called "fool's gold," shares little in common with gold other than the bright, metallic yellow color. Gold is highly malleable (bends easily) whereas pyrite is brittle and breaks easily. Streak color will also distinguish gold from pyrite.

WHERE TO LOOK: The Tombstone Hills, west of Tombstone, and the area around Morristown have mine dumps where you can find gold. Any river can also be searched for gold.

Pink, feldspar-rich granites

Coarse, "chunky" texture

Whiter, quartz-rich granite

Diorite specimen

Granite

HARDNESS: N/A **STREAK:** N/A

ENVIRONMENT: Mountains, roadcuts, desert, buttes, washes

Occurrence

WHAT TO LOOK FOR: Light-colored, coarse-grained rock with darker spots

SIZE: Ranges from small pebbles to entire mountains

COLOR: White to pink, with darker black or brown spots

OCCURRENCE: Very common

NOTES: Granite formed when molten rock cooled very slowly deep within the earth. This allowed the individual minerals in the rock to crystallize longer, allowing them to grow large enough to be visible. Other rocks, like basalt, cool at the earth's surface with very little time to crystallize, resulting in a dark rock with no visible crystals. Granite primarily contains feldspars and quartz, which contribute to its lighter hues. To simplify things, it may be helpful to consider the whiter granites as containing more quartz and the pinker granites as containing more feldspar. The darker spots within granite are primarily micas and pyroxenes, such as augite.

There are many similar coarse-grained rocks that greatly resemble granite, but their mineral composition is not of the characteristic granite type. One example is diorite, which contains more dark minerals. The term "granite," however, is applied to the entire group since distinguishing each individual type is often unnecessary and impractical, except for scientific purposes or for the serious rock collector.

WHERE TO LOOK: Granite is very common and can easily be found anywhere there is exposed rock. It will be less common in sandy areas, however, and more common in mountainous regions.

"Cauliflower-like" gypsum

Glauberite crystal replaced by gypsum

Sharp, angular points not characteristic of gypsum

Gypsum

HARDNESS: 1.5–2 **STREAK:** White

Occurrence

ENVIRONMENT: Mine dumps, roadcuts, desert, buttes

WHAT TO LOOK FOR: Flat, clear or white crystals that are easily scratched by your fingernails

SIZE: Gypsum and its varieties can be found in a wide range of sizes, from thumbnail-size up to basketball-size and larger

COLOR: Colorless to white or gray, also yellow and brown

OCCURRENCE: Common

NOTES: Gypsum is a very soft, common mineral that forms in cavities within sedimentary rock deposits, especially limestone. Its crystals are generally flat, which can be a good distinguishing characteristic, but they can also resemble cauliflower. Gypsum sometimes forms in rosettes, or thin blades of crystal arranged in a ball-like shape. Selenite, one of the many varieties of gypsum, most commonly forms these rosettes. Selenite is a particularly clear variety of gypsum that forms in mud or clay beds as thin, broad sheets of crystal. Gypsum often incorporates its surroundings, particularly sand, which can color its crystals in shades of brown to red. All varieties should be easily identified by its hardness alone—your fingernail will be enough to scratch it. Gypsum has long been used as the primary ingredient in plaster.

Glauberite, a rare colorless mineral related to salt, will actually turn into gypsum when exposed to air, becoming whiter and softer in hardness. This is known as a pseudomorph, or a mineral that has the appearance of a completely different mineral. This variety of gypsum is easy to identify, because glauberite has sharp, steeply pointed crystals.

WHERE TO LOOK: The area around Douglas is rich with gypsum, as well as the area around Sierra Vista.

Chrysocolla (blue)

Botryoidal hematite

Hematite coating on jasper

Hematite (black) in quartz (white)

Quartz

Hematite rosettes

Specimen courtesy of George Godas

Hematite

HARDNESS: 5–6 **STREAK:** Brownish red

Occurrence

ENVIRONMENT: Mine dumps, desert, buttes, mountains

WHAT TO LOOK FOR: Black or reddish metallic mineral

SIZE: Occurs in a wide range of sizes, from thumbnail-sized to basketball-sized and larger

COLOR: Steel-gray to black with brownish red oxidation

OCCURRENCE: Very common

NOTES: As the world's most important and common iron ore, hematite is very widespread throughout Arizona. It can occur in a number of forms but primarily appears botryoidally (in grape-like clusters) when well-formed. Tabular (flat) crystals are not uncommon and are often arranged into rosettes. However, you'll most likely find hematite within other rocks and minerals. Hematite is commonly found with quartz, calcite, barite and jasper. For example, jasper, a form of chalcedony, is often striped with alternating bands of hematite, and quartz is commonly stained red due to oxidized iron inclusions. In fact, hematite is so common that it would be difficult to find a region where hematite is not present in some form. It's easy to confuse hematite with goethite, another iron ore, because of its similar botryoidal structure. The easiest way to tell the two apart is to do a streak test. Hematite's streak is a reddish brown whereas goethite's will be yellow. Magnetite can also appear similar to hematite, but the easiest test in this case would be to apply a magnet to the specimens. A magnet will stick to magnetite but not to hematite.

WHERE TO LOOK: The Bradshaw Mountains south of Prescott, the Wickenburg Mountains northwest of Phoenix, and the mine dumps near Quartzsite all yield hematite.

Botryoidal hemimorphite

Specimens courtesy of George Godas

Large, clear crystals

Botryoidal specimens

Cross-section

Hemimorphite (gray) on aurichalcite (blue)

Hemimorphite

HARDNESS: 4.5–5 **STREAK:** Colorless

ENVIRONMENT: Mine dumps

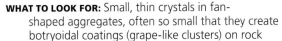

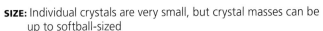

Occurrence

WHAT TO LOOK FOR: Small, thin crystals in fan-shaped aggregates, often so small that they create botryoidal coatings (grape-like clusters) on rock

SIZE: Individual crystals are very small, but crystal masses can be up to softball-sized

COLOR: Primarily colorless or white to gray, blue

OCCURRENCE: Uncommon

NOTES: Hemimorphite exhibits a unique and rare crystal form (habit). Its small, flat crystals are not symmetrical—one end of the crystal is flat while the other is pointed. Hemimorphite was named for this odd characteristic; its name translates roughly to "half-changed" in Greek. Even so, it's not always easy to observe this crystal structure, as the flat end of the crystal tends to be its base and is often attached to the host rock. In addition, most hemimorphite crystals are very small. They are normally colorless or white but are often colored by inclusions such as copper, which turns the mineral a beautiful blue color. When hematite inclusions are present, they tint the crystals brown. Hemimorphite is also commonly found with cerussite and galena. Crystals often form in fan-shaped aggregates when they are larger, but very small crystals tend to be very tightly packed together and appear as botryoidal crusts (grape-like clusters) on a matrix. These masses often have the appearance of velvet because each individual tiny crystal reflects light in a different direction. When broken, cross-sections of the botryoidal crusts sometimes exhibit attractive bands of white and blue that are used in jewelry.

WHERE TO LOOK: The Mammoth area mine dumps, the Dripping Spring Mountains, the Globe area and the Superior area.

Rough specimens

Orbicular jasper

Red and yellow banded jasper

Jasper

HARDNESS: 7 **STREAK:** White

ENVIRONMENT: Desert, buttes, washes, rivers, roadcuts

Occurrence

WHAT TO LOOK FOR: Hard, opaque masses in shades of red or brown with a waxy luster

SIZE: Jasper occurs massively and can be found in any size

COLOR: Brown, red, orange, yellow or green

OCCURRENCE: Common

NOTES: Jasper is essentially the more colorful equivalent of chert, since they both have the same mineral structure. Both consist primarily of quartz crystals that are so compact and small that the minerals become opaque and dense. While the white, gray and black varieties are more commonly called chert, jaspers are a wide range of colors. Red is the most common color variant that is produced by iron. In Arizona, red jasper is abundant and will sometimes be banded with other colors. And while some jasper forms in basalt or rhyolite vesicles (gas bubbles), most jasper is sedimentary and forms in flat layers at the bottom of bodies of water, often with alternating layers of hematite. Orbicular jasper is one of the many kinds of jasper and appears to be covered in "eyes." The circular jasper formations developed around quartz and feldspars that had developed earlier.

Jasper is easy to identify since it shares the same identifying characteristics as chert, chalcedony and quartz; all of these minerals are quite hard, have a waxy surface and fracture conchoidally (when struck, circular cracks appear). Jasper's lack of translucency rules out chalcedony, and its color variation (specifically reds) distinguishes it from chert.

WHERE TO LOOK: A huge area from Phoenix to Globe has produced jaspers, as have areas around Quartzsite.

Specimen courtesy of Mitchell Dale

Kaolinite

HARDNESS: 2–2.5 **STREAK:** White

ENVIRONMENT: Desert, buttes, mountains, washes

Occurrence

WHAT TO LOOK FOR: Soft, light-colored, layered earthy masses

SIZE: Kaolinite is massive (occurs in compact mineral concentrations) and can be any size

COLOR: White to gray, yellow to brown

OCCURRENCE: Common

NOTES: Kaolinite, like montmorillonite, is a common clay mineral. Its layered crystals are too small to see without a microscope and it occurs in layered masses of light-colored clay. Visually, it is nearly impossible to distinguish Arizona's kaolinite from similar minerals such as montmorillonite or bentonite. Kaolinite is slightly harder than montmorillonite, though both minerals' hardnesses can be somewhat variable and similar. One test that may or may not yield distinct results is the minerals' capacity to absorb water. Kaolinite absorbs considerably less water than montmorillonite or bentonite. Those two minerals will swell to several times their size when wet, whereas kaolinite will only swell slightly. If all practical tests fail to determine one mineral from the other, there is little more that you can do without professional analysis. Kaolinite is formed from the weathering of minerals rich in aluminum and silica (quartz), such as some feldspars. It can be found as exposed beds, which are easily "baked" in the sun and lose moisture, making the kaolinite more compact and dense. Despite its uninteresting appearance, kaolinite is an important mineral. It is used in the production of paper, as well as in ceramics because it doesn't shrink much when baked. It is also used in medicines for its aluminum content.

WHERE TO LOOK: Kaolinite is mostly found as layered deposits in desert areas where sedimentary material is found.

Kinoite

HARDNESS: 5 **STREAK:** Bluish white

ENVIRONMENT: Mine dumps

Occurrence

WHAT TO LOOK FOR: Small blue crystals, sometimes coating the surface of a rock

SIZE: Individual kinoite crystals are very small and no bigger than a pea, while kinoite coatings can be several inches in size

COLOR: Light to dark blue

OCCURRENCE: Rare

NOTES: Kinoite is quite rare and only occurs in a few locations in Arizona. One of many copper-based minerals, kinoite shares its bluish hues with similar minerals like azurite and chrysocolla, though hardness tests differentiate them. Well-crystallized kinoite is rare and it is mostly found as a thin coating or a vein of tiny crystals within its host rock.

It's not easy to confuse kinoite with very many other minerals, since its vivid blue color is distinctive. You're most likely to confuse kinoite with chalcanthite, though the two minerals crystallize in very different ways. Whereas kinoite's crystals remain small and form a coating on a rock, chalcanthite forms long, thin, wavy, hair-like crystals that are very brittle. In addition, chalcanthite is much softer.

WHERE TO LOOK: The Christmas mine, south of Globe, Arizona, is one of the state's best-known localities. The kinoite found there is unique in that it is almost always found with a thin coating of colorless or white apophyllite crystals on top of the kinoite crystals. In addition, the Dripping Spring Mountains, southeast of Phoenix, and the Sierrita Mountains, south of Tucson, all have mine dumps where kinoite can be found.

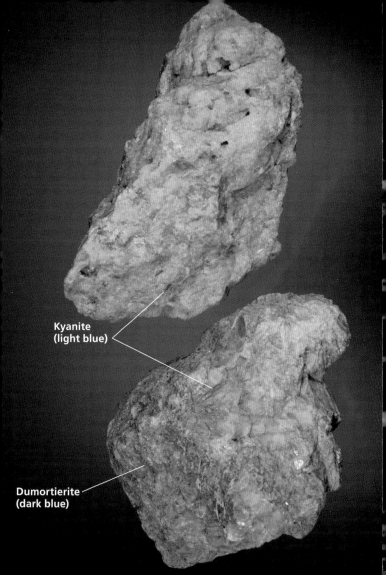

Kyanite
(light blue)

Dumortierite
(dark blue)

Kyanite

HARDNESS: 4–5 lengthwise, 6–7 crosswise

STREAK: Colorless

Occurrence

ENVIRONMENT: Washes, mountains

WHAT TO LOOK FOR: Blue crystals found embedded in other rock

SIZE: Kyanite crystals can be quite large but are generally golf ball-sized and smaller

COLOR: Blue to gray

OCCURRENCE: Uncommon

NOTES: Heat and pressure sometimes cause the various minerals within a rock to rearrange and compress; this process is called metamorphosis. Garnet is a result of this process, as is kyanite. Kyanite is the product of a metamorphosis of clay-rich rocks in which aluminum is present. Kyanite therefore is always found as embedded crystals, which can be hard to break free. The elongated, tabular (flat) crystals are usually colored in shades of blue, but gray is also common. As with other metamorphic minerals, kyanite can occur as very large crystals, sometimes over a foot long—but don't get your hopes up—most specimens are shorter than an inch long.

Kyanite has a unique identifying trait: it has two hardnesses. Lengthwise, or parallel to the crystal face, a knife will scratch the specimen, but crosswise it will not. Combined with the color, this characteristic is normally enough to identify the mineral.

WHERE TO LOOK: Most kyanite is found on the western border of the state. Granite Wash, near Quartzsite, Clip Wash, north of Yuma, and the Dome Rock Mountains, about 5 miles west of Quartzsite, have yielded kyanite.

Rough limestone

Fossil shell impression

Limestone

HARDNESS: 3–4 **STREAK:** N/A

Occurrence

ENVIRONMENT: Found in all environments

WHAT TO LOOK FOR: Abundant light-colored rock that consists primarily of fossil material

SIZE: Limestone can be found in any size—from tiny pebbles to entire mountains

COLOR: White to gray, darker when more fossil material is present, and more yellow or brown when iron is present

OCCURRENCE: Very common

NOTES: Limestone is an extremely prevalent sedimentary rock, especially in Arizona, and it would be hard not to pick up a piece. In order for a rock to be labelled limestone, its composition must be more than fifty percent calcite. The rest of the rock is normally made up of an assortment of other material, like quartz, dolomite or clays. Limestone is white or gray when pure but turns darker when there is more fossil material present. Similarly, it turns more yellow or brown when iron is present. The easiest test to determine if a specimen is limestone is to apply a drop of vinegar. True limestone will bubble and fizz in the acid. Since limestone is a sedimentary rock, it commonly preserves ancient animals within its structure. Small, hard animals like snails, clams and corals have their shells embedded in the rock, and, in fact, there are varieties of limestone that consist nearly entirely of fossil material. Chalk, a bright white variety, is composed primarily of fossilized algae and other microscopic animals but is uncommon in Arizona. Travertine is a variety of layered (and sometimes colorful) limestone that is found most commonly in caves and around hot springs.

WHERE TO LOOK: Limestone is extremely abundant and can be found everywhere. The map indicates the most prominent limestone formations.

Orange-brown coloration

Limonite replacement of cubic pyrite crystals

Limonite replacement of cubic pyrite crystal

Limonite

HARDNESS: 4–5.5 **STREAK:** Yellowish brown

ENVIRONMENT: Mine dumps, desert, buttes, mountains

Occurrence

WHAT TO LOOK FOR: Brown masses of structureless material that often forms on goethite

SIZE: Limonite occurs massively (in compact crystal concentrations) and can be found in any size

COLOR: Brown to orange-brown, rust-colored, black

OCCURRENCE: Very common

NOTES: If you've ever seen a rust-colored patch of soil, brownish yellow coloration in jasper, or a dusty orange-brown coating on a rock, it was probably limonite. The name limonite refers to any earthy, rusty brown, otherwise unidentified iron oxide that contains water. In essence, limonite is the same thing as goethite, but where goethite forms fibrous crystals, limonite occurs as massive, soil-like chunks with no crystal structure. It is a result of iron oxides weathering and decomposing. And since Arizona is so rich in iron oxides, especially goethite, you're likely to find limonite virtually anywhere you look. Limonite has no crystal structure of its own but is often found replacing other minerals' crystals, primarily pyrite. Since limonite and pyrite occur in many of the same mineral environments, limonite will often coat and then replace pyrite cubes—sometimes so delicately that every detail of the original pyrite is preserved, including surface striations (fine grooves). Limonite is fairly distinctive and you're unlikely to confuse it with other minerals. Goethite, while occurring in the same colors, has a fibrous cross-section when broken and hematite has a reddish streak.

WHERE TO LOOK: The Globe and Miami areas have yielded nice specimens, as have the Patagonia and Huachuca Mountains.

Specimen courtesy of Mitchell Dale

Rock matrix

Magnesite (white)

Polished cabochon

Courtesy of Mitchell Dale

Chalky, massive magnesite

Jasper (red)

Polished magnesite crystals (yellow/white)

Magnesite

HARDNESS: 3.5–4.5 **STREAK:** White

Occurrence

ENVIRONMENT: Buttes, roadcuts, mountains, desert

WHAT TO LOOK FOR: Light-colored, chalky masses or crystals, often embedded within rock

SIZE: Embedded magnesite crystals are generally thumbnail-sized or smaller, while massive magnesite (compact mineral concentrations) can be larger

COLOR: White to yellow, gray, or brown

OCCURRENCE: Uncommon

NOTES: Magnesite is a light-colored mineral named for its magnesium content. Colorless or yellow magnesite crystals are beautiful but very rare in Arizona; magnesite is mostly found as massive chunks or as pockets within rock. Such specimens are often featureless with a rough, earthy and slightly chalky texture and offer little to collectors. However, as it is generally quite solid, it can be cut and polished; such specimens are sometimes used in jewelry. On the other hand, magnesite that occurs as cauliflower-shaped pockets or veins in rock are quite collectible. Magnesite found this way sometimes stands up off of the surface of the surround-ing rock, which weathers faster. Such hard masses often have a porcelain-like luster and polish well. "Wild Horse magnesite," a variety with a reddish brown matrix, is well known in Arizona. Magnesite forms as a result of serpentine that weathers and breaks down, so magnesite can some-times be found alongside green serpentines, though confusing the two is unlikely. Magnesite can be confused, however, with calcite and dolomite, though both are softer.

WHERE TO LOOK: The Cave Creek area, north of Phoenix, and the Bisbee area are both known Arizona localities for magnesite.

Embedded octahedral crystals

Limonite coating (brown)

Loose octahedral crystals

Magnetite crystals

Calcite

Magnetite

HARDNESS: 5.5–6.5 **STREAK:** Black

ENVIRONMENT: Mine dumps, mountains

WHAT TO LOOK FOR: Black, metallic and magnetic mineral

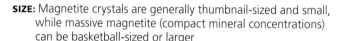

Occurrence

SIZE: Magnetite crystals are generally thumbnail-sized and small, while massive magnetite (compact mineral concentrations) can be basketball-sized or larger

COLOR: Iron-black

OCCURRENCE: Common

NOTES: Magnetite is a widespread iron mineral and, while not as common as hematite, is easy to find in well-mineralized (mineralogically diverse) regions. The dark, metallic mineral can occur massively as nondescript black chunks, but its octahedral (eight-faced) crystals are common and much more collectible. Its most unique characteristic is that it attracts a magnet, which no other native Arizona mineral does. (Meteorites also are magnetic.) A magnet therefore is the best test for magnetite, as it should easily stick to the mineral.

Hematite and goethite can be confused with magnetite, but magnetite's dark black streak and magnetism easily set it apart. Magnetite occurs with other iron oxides, particularly limonite, with which it is never easily confused. It also can form embedded within other minerals, such as calcite.

A variety of magnetite known as lodestone is naturally magnetic and will attract small pieces of iron, as well as magnetite itself.

WHERE TO LOOK: Mine dumps in the Patagonia Mountains and the Santa Rita Mountains, south of Tucson, as well as several mines closer to the Mexican border, have produced magnetite.

Limonite (orange)

Banded malachite

Azurite (blue)

Tenorite (black)

Botryoidal malachite

Azurite (blue)

Malachite

HARDNESS: 3.5–4 **STREAK:** Light green

Occurrence

ENVIRONMENT: Mine dumps, desert, buttes

WHAT TO LOOK FOR: Green crystals or banded masses on or in a host rock, often occurring with blue azurite

SIZE: Crystals are generally thumbnail-sized and smaller; massive malachite (compact mineral concentrations) can be up to softball-sized and rarely larger

COLOR: Green to dark green

OCCURRENCE: Uncommon

NOTES: Malachite is one of Arizona's best known collectibles and the state has produced some of North America's finest specimens. It is highly sought after for its rich green color and its banded cross-section, which is frequently used in jewelry. Individual crystals are much rarer than massive pieces or botryoidal (grape-like) crusts found atop other minerals.

More often than not, malachite is associated with azurite, its blue mineral cousin, and they often occur in the same specimens. This is because malachite and azurite have virtually the same chemical composition. In fact, large masses of malachite sometimes have a core made of azurite. This trait, combined with malachite's characteristic color, are good identifiers. Malachite is also found as thin, dusty green films on copper and other copper-related minerals. Therefore, copper miners have used malachite as an indicator of where to find copper ore.

Malachite occurs with azurite, limonite, chalcopyrite and copper.

WHERE TO LOOK: Arizona's best malachite has come from the Globe, Bisbee, Superior, Jerome and Quartzsite area mine dumps.

Green marble, colored by mica inclusions

Pink marble

Layered marble

Marble

HARDNESS: 3 **STREAK:** N/A

ENVIRONMENT: Mountains, buttes, roadcuts

WHAT TO LOOK FOR: Soft, light-colored rock in limestone-rich areas

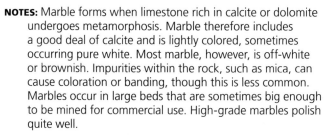
Occurrence

SIZE: Marble occurs massively and can be found in any size

COLOR: White or gray, yellow to brown, and rarely pink or green

OCCURRENCE: Common

NOTES: Marble forms when limestone rich in calcite or dolomite undergoes metamorphosis. Marble therefore includes a good deal of calcite and is lightly colored, sometimes occurring pure white. Most marble, however, is off-white or brownish. Impurities within the rock, such as mica, can cause coloration or banding, though this is less common. Marbles occur in large beds that are sometimes big enough to be mined for commercial use. High-grade marbles polish quite well.

Gneisses and schists are sometimes found with marble because of the similar metamorphic events that create the rocks, though confusing them with marble is unlikely. Silica (quartz) sometimes inundates very old marble deposits, making it much harder—about a 5.5 on the Mohs hardness scale. But you'll most likely come across the softer, more common type, which is easy to identify. Marble exhibits most of the testable traits of calcite or the limestone it derives from. It will be scratched by a copper coin as well as effervesce in acid. In addition, limestone sometimes contains fossils (e.g. shells); marble can too, though rarely.

WHERE TO LOOK: Marble is found in several places, including the area around Quartzsite and the Santa Rita Mountains, south of Tucson, as well as the Dragoon Mountains, north of Tombstone.

Specimens of meteorite fragments fused with rock

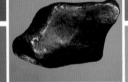

Partially polished meteorite

Sharp, angular shape

Smooth, rusty surface

Meteorites

HARDNESS: N/A   **STREAK:** N/A

ENVIRONMENT: Desert, buttes, washes, rivers

WHAT TO LOOK FOR: Dark, heavy pieces of metallic
material

Occurrence

SIZE: Meteorites occur in a wide range of sizes but are generally
smaller than your fist

COLOR: Gray to black, brown, rust-colored

OCCURRENCE: Very rare

NOTES: Meteors are rocks that fall from space; most of them
burn up in the atmosphere. When this happens, we see a
"shooting star." Sometimes meteors impact the surface and
can be found as a meteorite. While there are several types
of meteorites, most are hard, extremely heavy mixtures of
nickel and iron. Pure iron is very rare on earth, as it oxidizes
when exposed to oxygen; meteorites are one of the few
natural sources of pure iron. Most meteorites are found
fused with other rock. This is due to the tremendous heat
and energy that occurs during an impact, melting the
surrounding rock. Arizona is home to the famous Meteor
Crater (sometimes known as Barringer Crater or Canyon
Diablo), the world's best-preserved meteor impact crater.
Nearly a mile wide, it formed when a meteor 160 feet
across hit the ground, exploding and hurling fragments for
hundreds of miles. In fact, most meteorites found in Arizona
are considered to be fragments of this original impact.
When hunting for meteorites, look for rocks with a fusion
crust (a layer of metallic material that is melted and fused
with rock). Be sure to bring a magnet; most meteorites
attract a magnet because of their iron content. Meteorites
won't normally be confused with anything except magnetite, which is also magnetic.

WHERE TO LOOK: There are no definite locations for meteorites
due to their scattered and random nature.

Mass of intergrown muscovite crystals

Compact mica crystals

High luster

Mica "book"

Mica Group

HARDNESS: 2–3 **STREAK:** Colorless

ENVIRONMENT: Mine dumps, mountains, roadcuts

WHAT TO LOOK FOR: Thin, flaky crystals with a high luster that are arranged in stacks

Occurrence

SIZE: Individual mica crystals can occur quite large, but are generally thumbnail-sized and smaller

COLOR: Brown to greenish brown, black

OCCURRENCE: Common

NOTES: The term "mica" refers to an important group of minerals rather than just one specific mineral. These distinct minerals are all soft and occur as thin, flaky crystals that form in "books" or stacks. Individual sheet-like crystals can be peeled back and separated from the rest of the stack. Loose crystals of mica are flexible and can bend quite a bit before breaking. They are also commonly transparent and have a high luster.

Micas are most common within rocks. Some granites have dark, shiny spots in them that are a type of mica called biotite. Metamorphic schists also can contain lots of mica, and one specific type—mica schist—consists almost entirely of tiny pieces of mica arranged in parallel layers.

The three most common types of mica are biotite, phlogopite and muscovite, all of which are easy to identify as micas, but they can be difficult to differentiate. Biotite is generally dark shades of gray or black and has a high luster, muscovite is more brown or green-colored, and phlogopite is darker than muscovite, but not as dark as biotite.

WHERE TO LOOK: Fine specimens have come from the Dome Rock Mountains, about 5 miles west of Quartzsite.

Mimetite

Hemimorphite

Mimetite (orange)

Globular mimetite on barite

Needle-like crystals

Mimetite

HARDNESS: 3.5–4 **STREAK:** White

ENVIRONMENT: Mine dumps

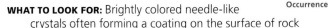

Occurrence

WHAT TO LOOK FOR: Brightly colored needle-like crystals often forming a coating on the surface of rock

SIZE: Generally very small crystals that measure just millimeters long

COLOR: Yellow, orange, brown or colorless

OCCURRENCE: Rare

NOTES: Mimetite is a rare lead-based mineral that forms small, attractive crystals. Mimetite comes in a variety of colors, but Arizona is well-known for richly colored orange specimens. Crystals are slender, needle-like and hexagonal (six-sided) and often grow together in intricate masses, though less commonly, crystals can occur as globules (ball-like crystals), too. These crystals can be found in cavities within rocks as well as growing on the surface of other minerals.

As a lead mineral, mimetite is commonly found with galena, the primary ore of lead. Mimetite also grows with, or on, barite. Attractive specimens of mimetite perched on the tip of barite blades have been found in Arizona, but are very rare. Also, in at least one locality, mimetite has been found in vugs (cavities) within masses of hemimorphite.

Mimetite and vanadinite are closely related and can possibly be confused, though their crystal shapes are often enough to differentiate the two. Mimetite is also much rarer than vanadinite.

WHERE TO LOOK: The mine dumps in the Bisbee and Warren areas, as well as the Superior and Mammoth areas, are all well known for mimetite.

Hexagonal crystal shape

Quartz

Flat, hexagonal crystal

Molybdenite flakes in rock

Molybdenite

HARDNESS: 1–1.5 **STREAK:** Gray green to black

ENVIRONMENT: Mine dumps, mountains

Occurrence

WHAT TO LOOK FOR: Soft, bluish metallic crystals formed of hexagonal layers on quartz

SIZE: Individual crystals are thumbnail-sized or smaller, and flakes found within other rocks can be very small

COLOR: Bluish lead-gray

OCCURRENCE: Uncommon

NOTES: Molybdenite is named for its high content of molybdenum and is one of the primary sources of the element. Its metallic, bluish, hexagonal (six-sided) crystals are normally found no larger than your thumbnail and appear to be layered. Due to molybdenite's low hardness and flexibility, these scaly layers are often misshapen or deformed because of the surrounding matrix pressing on the crystals. The mineral's color and crystal shape are generally enough to identify it, as few other minerals combine similar traits. Its streak color also helps to identify it—it tends to be a dark gray or black on a piece of white paper and dark green on a streak plate.

Molybdenite can also occur as flakes embedded within rock. These specimens tend to be less collectible than finely shaped crystals. Molybdenite can often be confused with various micas due to its dark color and layered habit, but hardness and streak, as well as the minerals with which it occurs, differentiate it. Molybdenite occurs with quartz, chalcopyrite, pyrite, garnet and barite.

WHERE TO LOOK: Many of the mines in the Patagonia Mountains, north of Nogales, have produced molybdenite and you may be able to find some in mine dumps in the area.

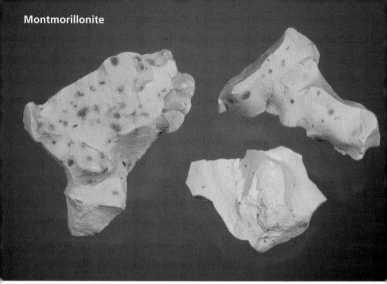

Montmorillonite

Bentonite

Montmorillonite

HARDNESS: 1–2 **STREAK:** White

ENVIRONMENT: Desert, buttes, roadcuts

Occurrence

WHAT TO LOOK FOR: Soft, lightweight, soapy-feeling clay

SIZE: Montmorillonite occurs massively and can be found in any size

COLOR: White to gray, yellow to brown, flesh-colored

OCCURRENCE: Very common

NOTES: Montmorillonite is a very common clay mineral in Arizona. Its crystals are thin plates that are packed very tightly together and can only be seen on a microscopic level. The mineral is found as large beds of whitish, soft, lightweight clay that has a slightly soapy feel. The microscopic, layered habit of its crystals make the mineral very porous and absorbent—when wet, a specimen of montmorillonite can expand up to ten times its original size. However, this also causes montmorillonite to weather easily and crumble. Though interesting, it isn't a particularly collectible or valuable mineral. Montmorillonite is derived from ancient beds of volcanic ash and tuff that have weathered. An impure variety of montmorillonite is bentonite, which contains more ash and water, and occurs in a wide variety of colors, from reds and oranges to blues. Arizona's Painted Desert, known for its multi-colored hills, consists primarily of bentonite clays. Another mineral closely related to montmorillonite is saponite, which is a common vesicular mineral in Arizona. Montmorillonite minerals are made into a slurry (mud) that is used as a lubricant for drills. They are also used in cements and ceramics, as well as in cat litter, due to their absorbent nature.

WHERE TO LOOK: Montmorillonite, specifically bentonite, is very abundant, especially in the northeast corner of the state.

Silky luster

Fibrous texture

Specimen courtesy of Mitchell Dale

⚠ Mountain Leather (Asbestos)

HARDNESS: ~2–3 **STREAK:** N/A

ENVIRONMENT: Desert, buttes, mountain

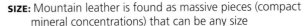

Occurrence

WHAT TO LOOK FOR: Fibrous "woven" mats of soft material

SIZE: Mountain leather is found as massive pieces (compact mineral concentrations) that can be any size

COLOR: White to gray, yellow to brown

OCCURRENCE: Uncommon

NOTES: "Mountain leather" is a nickname given to a very unique mineral growth. Also known as "mountain cork" or "desert leather," mountain leather is actually one of two asbestos minerals: tremolite or chrysotile serpentine. Most mountain leather, however, is tremolite, a light-colored mineral. It forms soft, fibrous mats of asbestos that appear to be woven together. Thin samples are flexible, strong and fabric-like and are often found following the contours of the rock beneath it. Mountain leather is soft and can have a wavy, flowing surface that sometimes has a silken luster. The term asbestos doesn't pertain to any specific mineral. Instead, it refers to any light, flexible, fibrous minerals that appear silky and are often fire-resistant. The fibers of asbestos minerals can actually be woven into fabrics and were historically used as fireproof insulation material. However, asbestos has gained notoriety in recent years, as we now know asbestos inhalation causes illness, specifically cancer. Always handle asbestos minerals, such as tremolite, serpentine and mountain leather, with care and store them in airtight boxes.

WHERE TO LOOK: The area around Globe has yielded some mountain leather as well as Ash Peak, which is about halfway between Safford and Duncan.

"Apache tears"

Perlite matrix

Round, glassy balls

Mud coating

Black, glassy surface

Conchoidal (round) fracture

Obsidian

HARDNESS: 6–7 **STREAK:** N/A

ENVIRONMENT: Desert, buttes, washes

WHAT TO LOOK FOR: Dark, clear rock greatly resembling glass

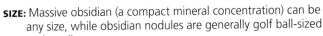

Occurrence

SIZE: Massive obsidian (a compact mineral concentration) can be any size, while obsidian nodules are generally golf ball-sized and smaller

COLOR: Black to transparent gray

OCCURRENCE: Common

NOTES: Obsidian is frequently referred to as "volcanic glass," which is an apt description. Obsidian formed when volcanic lava flows cooled very rapidly, such as when they erupted into a body of water, or when a lava flow was extremely thick and nearly devoid of water. In either case, the individual minerals within the rock were not allowed to crystallize and the result is a blank, dark, transparent rock with no visible crystals. If the molten material had been allowed to cool normally, it would have formed a light-colored granite.

Obsidian contains mostly feldspars but also has a lot of quartz within it, which makes it hard and gives it a conchoidal (circular) fracture. Ancient Americans often used obsidian in their tools and weapons because its hardness and glass-like nature produced a razor-sharp cutting edge. Arizona is home to a particular type of obsidian known locally as "Apache tears." These are round, nodular obsidians that form from larger bodies of obsidian that have become altered by water and heat. The altered obsidian is called perlite, and it contains more water than true obsidian. All types of obsidian are difficult to confuse with other minerals.

WHERE TO LOOK: The San Francisco Mountains, north of Flagstaff, and the San Francisco River, south of Clifton, are known sites for obsidian.

Peridot grains

Polished peridot

Peridot grains in basalt matrix

Olivine

HARDNESS: 6.5–7 **STREAK:** Colorless

ENVIRONMENT: Desert, buttes, roadcuts, mountains

WHAT TO LOOK FOR: Clear green grains found embedded in basalt

SIZE: Olivine is normally found as small, pea-sized grains

COLOR: Yellowish green to brown

OCCURRENCE: Uncommon

NOTES: Olivine is a fairly common mineral that acts as an important rock-building component in dark rocks like basalt and gabbro. It occurs in shades of green and brown and is generally not found as individual crystals, but rather as granular masses. Transparent, gem-quality olivine is called peridot and is one of Arizona's well-known collectibles. Peridot is much more rare than standard olivine and does not occur in very large pieces; yet good, clear specimens are often used in jewelry. A deep, rich olive-green color is considered the most valuable.

In Arizona, peridot is found as grains within pockets in basalt. Loose grains may occasionally be confused with quartz because of their similar hardness, but quartz rarely appears in the same vivid green color. In addition, peridot is far more rare than quartz and in Arizona, nearly all the available peridot occurs within the San Carlos Native American Reservation and is not accessible to the public. Residents of the reservation mine the mineral and sell specimens as well as peridot jewelry. Peridot is also the traditional birthstone for the month of August.

WHERE TO LOOK: Olivine can be collected in the Patagonia Mountains and the Santa Rita Mountains, north of Nogales, as well as in the San Francisco Volcanic Field, north of Flagstaff.

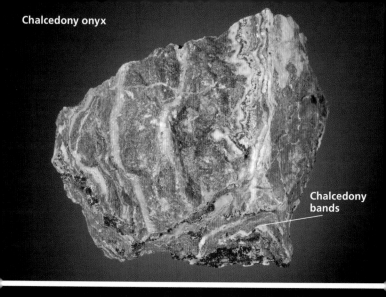

Chalcedony onyx

Chalcedony bands

Calcite bands

Onyx marble

Onyx

HARDNESS: N/A **STREAK:** White

ENVIRONMENT: Desert, buttes, roadcuts

WHAT TO LOOK FOR: Multi-layered minerals

SIZE: Most onyx is massive (occurs in compact mineral concentrations) and can be any size

COLOR: Varies greatly

OCCURRENCE: Uncommon

NOTES: The name onyx is more of a descriptive term rather than any particular mineral. While old definitions state that true onyx is a layered form of chalcedony, other minerals, such as calcite or marble, can be considered onyx when they also form in parallel bands. Hardness tests are best in determining calcite-based onyxes from chalcedony onyx.

Chalcedony onyx differs from agate in that it formed in parallel layers that go across and through the material, as mentioned above. Agates form in concentric bands, or circular layers that surround each other like a bull's-eye pattern. Agates also occur in the vesicles (gas bubbles) within basalt, whereas chalcedony onyx forms massively. Calcite onyx, or onyx marble, is often found in shades of yellow or brown and is much softer than chalcedony onyx. A lot of onyx marble is found hanging from the ceiling of limestone caves as stalactites. Aside from stalactites, onyx forms when calcite-rich liquid solutions dry out and settle into layers. A third variety of onyx, called travertine, is often deposited around natural hot springs. Travertine is composed primarily of aragonite, a close mineral cousin to calcite, and it is often considered a limestone.

WHERE TO LOOK: Stockton Hill, northwest of Kingman, and the Bradshaw Mountains, south of Prescott, are good places to start.

Granular pyrite

Massive pyrite

Pyritohedron

Pyrite

Chalcopyrite

Pyrite

HARDNESS: 6–6.5 **STREAK:** Greenish black

ENVIRONMENT: Mine dumps, mountains, buttes, roadcuts

Occurrence

WHAT TO LOOK FOR: Golden, metallic masses with a cubic nature

SIZE: Pyrite crystals are generally thumbnail-sized and smaller, while massive pyrite (which occurs in compact mineral concentrations) can be much larger

COLOR: Metallic yellow; brown if tarnished

OCCURRENCE: Common

NOTES: Known for decades as "fool's gold," pyrite's golden yellow color has built up the hopes of more than one collector. A combination of iron and sulfur, it is one of the most common iron minerals and features beautiful, shiny, metallic, yellow crystals. It differs greatly from gold, however, due to its much greater hardness, brittle nature and great abundance. Pyrite crystals are very often cubic, but also can occur in unique pyritohedrons—a shape named for pyrite. Pyritohedrons are soccer ball-like crystals that consist of twelve pentagonal (five-sided) faces. Pyrite also occurs massively and in granular form and can be found as chunks or flakes. Whatever the variety, pyrite is a well-known and sought after collectible.

Pyrite is extremely widespread and is found in nearly any iron deposit. It occurs in numerous mineral environments with many other minerals, such as quartz, feldspars, micas and chalcopyrite. Pyrite can easily be confused with chalcopyrite, though chalcopyrite is much softer and often has a colorful, iridescent tarnish. Pyrite also forms alongside limonite, which can sometimes coat or replace pyrite.

WHERE TO LOOK: Mine dumps in Superior, Miami and Globe, as well as mines in the mountains south of those cities, are all well known for pyrite.

Dull gray metallic color

Finer-quality crystals

Massive habit (form)

Fibrous, radiating structure

Pyrolusite

HARDNESS: 6–6.5 **STREAK:** Sooty black

Occurrence

ENVIRONMENT: Mine dumps, mountains, desert, buttes

WHAT TO LOOK FOR: Dark, fibrous, metallic masses or crystals

SIZE: Massive pyrolusite (occurring in compact mineral concentrations) can be any size, but crystals tend to be thumbnail-sized or smaller

COLOR: Steel-gray to black

OCCURRENCE: Common

NOTES: Pyrolusite is a variety of oxidized manganese and is the most abundant manganese mineral. While it most commonly forms in dark indistinct masses, pyrolusite sometimes exhibits a fibrous, radiating crystal structure. Very rarely, pyrolusite can be found as small tabular (flat) crystals growing in a cavity, but don't get your hopes up, as this is unlikely. Pyrolusite also occurs as dendrites (tree-like growths) forming on the face of limestones; in chalcedony, these pyrolusite formations are called "moss agate."

Dark manganese-based minerals like pyrolusite have many traits in common with one another. A black, sooty (dark powdery) streak is very characteristic of manganese minerals and some are so powdery that they will blacken your hands when held. With that said, no trait is distinctive enough to tell them apart. In fact, x-ray tests are nearly the only way to tell various black, massive manganese minerals from one another. Therefore, the name pyrolusite is considered to be a safe label for most dark, sooty, massive minerals or for unidentified fibrous, metallic crystals that have a black powdery streak.

WHERE TO LOOK: The mine dumps in the hills west of Tombstone as well as the mines near Courtland and Gleeson in the Dragoon Mountains.

Mass of intergrown crystals

Drusy quartz

Quartz points

Quartz stained green
by chrysocolla

Amethyst point

Amethyst

Quartz

HARDNESS: 7 **STREAK:** White

ENVIRONMENT: Found in all environments

WHAT TO LOOK FOR: Light-colored, hard, glassy crystals or masses

Occurrence

SIZE: Quartz varies greatly in size, from tiny pebbles to foot-long crystal points

COLOR: Colorless, white to gray, yellow to brown, as well as blue, green, purple or pink

OCCURRENCE: Very common

NOTES: Quartz is the single most abundant mineral on the planet. Composed of silicon and oxygen (also known as silica), quartz is a very hard mineral that occurs in an incredible variety of forms, from large crystal points to chalcedony, agate, chert and jasper, which are all composed nearly entirely of microcrystalline quartz (microscopic quartz crystals). Quartz has a glassy luster, exhibits conchoidal fracture (when struck, circular cracks form) and is harder than the majority of other minerals, so identifying quartz is usually easy. Quartz crystals are six-sided and often form beautiful points, which are sometimes called "rock crystal." Most quartz, however, won't be found in well-formed points. It is most commonly incorporated into rocks such as granite. In fact, quartz is probably present in any hard, light-colored rock you find. When small quartz crystals coat the surface of a rock or a mineral, it is referred to as drusy quartz. There are many color variants of quartz, which makes it a diverse collectible. Purple quartz is called amethyst, pink is rose quartz, yellow is citrine, gray or black is smoky quartz, and greenish or bluish quartz often has micas or chrysocolla within it.

WHERE TO LOOK: Quartz is extremely common and can be found virtually everywhere.

Grainy texture

Quartz-like appearance

Quartzite

HARDNESS: ~7 **STREAK:** N/A

Occurrence

ENVIRONMENT: Desert, buttes, rivers, washes, roadcuts

WHAT TO LOOK FOR: Light-colored, grainy, hard rock

SIZE: Quartzite can occur in any size

COLOR: White to light brown, also pale pink or yellow

OCCURRENCE: Common

NOTES: Sandstone's primary ingredient is silica sand (sand that consists of quartz). Sometimes a silica solution will cement these grains together and form a much harder, denser version of sandstone called quartzite. Whereas sandstone is often crumbly and you can separate individual grains from the rock, quartzite is much more solid and less porous. As a result, sandstone breaks around its sand grains whereas quartzite will break right through them, though this trait is not easily observable.

In reality, you're very unlikely to get sandstone confused with quartzite. It is more likely, however, that you'll confuse it with pure quartz. A hardness test won't be much help since quartzite contains so much silica that its hardness is about the same as quartz's, although sometimes quartzite is slightly softer. A microscope or hand lens is the best tool to use for identification because where white, massive quartz tends to have a smooth face, quartzite appears grainy. Sometimes the individual grains are hard to see, but in general, you should be able to tell if it was once sandstone.

WHERE TO LOOK: Quartzite is quite abundant in Arizona and can be found in virtually any desert area.

Red fine-grained rhyolite

Flow structure

Polished cabochon

Fine-grained pumice

"Frothy" texture
of pumice

Orbicular rhyolite

Green-banded rhyolite

Rhyolite

HARDNESS: 6–6.5 **STREAK:** N/A

Occurrence

ENVIRONMENT: Washes, mountains, roadcuts, buttes

WHAT TO LOOK FOR: Hard, reddish, fine-grained rock

SIZE: Rhyolite is massive (it occurs in compact concentrations) and can be any size

COLOR: Pink to red or brownish red, white to gray, yellow to brown

OCCURRENCE: Very common

NOTES: Rhyolite, like basalt, is a fine-grained volcanic rock that cooled once it reached the earth's surface, but it is harder and more lightly colored than basalt because of its higher quartz content. Rhyolite, in its molten form, is more viscous than basalt, meaning that it is thicker and less "runny," which results in more gases being trapped within the rock. Some rhyolite is actually so full of vesicles (gas bubbles) that it resembles the foam that forms atop soft drinks. This "frothy" rhyolite is called pumice and is so full of air that even large boulders weigh very little and some specimens will even float on water! Rhyolite will often exhibit a flow structure, visible as bands or stripes in the rock. This banding shows that the molten rhyolite was moving when it cooled. Rhyolite can also form orbicularly, or circularly. Collectible rhyolite tends to be of the banded or orbicular variety. And since it is so hard, it can even be polished and used in jewelry. Rhyolite's mineral composition is the same as granite, but molten granite had a longer amount of time to cool and formed larger crystals and grains within the rock.

WHERE TO LOOK: Rhyolite is extremely prevalent, but the hills and washes in the Quartzsite area are good places to find the banded varieties.

Rosasite (blue)

"Puff ball" aggregates

Rosasite

HARDNESS: 4 **STREAK:** Light blue

Occurrence

ENVIRONMENT: Mine dumps

WHAT TO LOOK FOR: Small, round, blue and "fuzzy" crystals on the surface of rock

SIZE: Rosasite is found very small and is often just a millimeter or two in size

COLOR: Light blue to dark blue, teal

OCCURRENCE: Uncommon

NOTES: Rosasite is a mixture of copper and zinc and is very similar to aurichalcite. In fact, the two minerals can appear nearly identical at times. Rosasite is harder, though doing a scratch test on crystals that don't get much more than a millimeter in size is very difficult. Instead, you'll have to rely on observation. Rosasite's crystals tend to form botryoidal (grape-like) crusts or soft-looking radial "puff ball" bunches that are quite small and often grow on limonite. Aurichalcite's crystals also occur in radiating aggregates but are often longer and less tightly grouped.

Rosasite is quite difficult to find due to its rarity and small crystal size. Check vugs (cavities) in rocks at sites with copper deposits for the characteristic blue-green color. Some rosasites from Arizona have been found coated in a layer of calcite, which gives the normally "fuzzy" crystal groupings a darker, glassy, "wet" look.

WHERE TO LOOK: Rosasite has been known to come from mine dumps in the Tombstone Hills, west of Tombstone, the Gleeson and Courtland areas in the Dragoon Mountains, and in the Chiricahua Mountains, near Sierra Vista.

Dark red rutile crystal

Quartz

Muscovite mica

Rutile

Rutile

HARDNESS: 6–6.5 **STREAK:** White-brown

ENVIRONMENT: Mountains, mine dumps, roadcuts

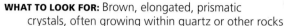
Occurrence

WHAT TO LOOK FOR: Brown, elongated, prismatic crystals, often growing within quartz or other rocks

SIZE: Generally, rutile crystals are small and thin, no longer than an inch or two long

COLOR: Reddish black normally, but brown or gold when oxidized

OCCURRENCE: Uncommon

NOTES: Rutile is an easily identified mineral that is often found as complete crystals. While its primary ingredient is titanium, rutile is more common than one might think and is often found as small, embedded crystals within metamorphic rocks like schists or gneisses, and, rarely, in other minerals, especially quartz. These crystals are frequently long and thin. Its color depends on how oxidized (combined with oxygen) its titanium content has become. Fresh rutile is reddish black while specimens that have been exposed to air longer turn golden yellow or brown.

Rutile's colors, crystal habit (form), and its occurrence within quartz or metamorphic rocks are all good identifiers for the mineral. In addition, well-formed groups of rutile crystals form at perfect 60 degree angles to each other, creating what are described as "knee-like" aggregates, though these are rare. Arizona has produced fine, red-colored rutile formed in broad crystals occurring with quartz.

WHERE TO LOOK: Mines in the Bisbee and Warren area have produced rutile specimens, as have the Patagonia Mountains, north of Nogales. Mine dumps in the Globe area have also produced rutile.

Crumbly, weathered surface

Differently colored sand

Granular surface texture

Finely banded sandstone

Sandstone

HARDNESS: N/A **STREAK:** N/A

ENVIRONMENT: Desert, buttes, roadcuts, washes

WHAT TO LOOK FOR: Red or yellow grainy, layered rocks

Occurrence

SIZE: Sandstone occurs in massive beds (compact concentrations) and can be found in any size

COLOR: White to yellow or brown, pink to red

OCCURRENCE: Very common

NOTES: Sandstone is exactly what it sounds like—rock made from sand. This sedimentary rock forms when beds of sand are compacted and cemented together by finer material, like clay or silt. Since sand is primarily quartz, you might think sandstone should be quite hard, when in reality most sandstones are very poorly compacted. Different varieties of sandstone are compacted differently and hardness tests therefore don't mean much. Loosely formed sandstone crumbles easily and you can pull individual grains of sand off of the rock. Sandstone that is cemented together by silica (quartz) is much harder and denser but is called quartzite.

While sandstone can be white or yellow, it is often found in shades of red or brown, as it is commonly stained by hematite and other iron oxides. Rarer, more exciting, varieties of sandstone exhibit twisted, wavy layers of different colors caused by millions of years of being pressed and pushed around by other rocks.

Arizona is rich with sandstone and many cliffs, hills and ridges are made of it. It should be quite easy to identify, but the best test is its feel. Sandstone nearly always feels gritty and rough, like sand.

WHERE TO LOOK: Sandstone can easily be found throughout the entire state.

Saponite lining vesicles (gas bubbles)

Basalt matrix

Analcime crystals (white)

Basalt vesicle (gas bubble) filled with saponite

Saponite

HARDNESS: 1.5–2 **STREAK:** White

ENVIRONMENT: Mine dumps, mountains, buttes

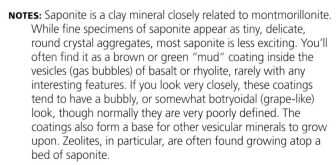

Occurrence

WHAT TO LOOK FOR: Light-colored, very soft material lining cavities in basalt or rhyolite

SIZE: Saponite is usually found as coatings within vesicles (gas bubbles)

COLOR: White to gray, brown, green, yellow

OCCURRENCE: Common

NOTES: Saponite is a clay mineral closely related to montmorillonite. While fine specimens of saponite appear as tiny, delicate, round crystal aggregates, most saponite is less exciting. You'll often find it as a brown or green "mud" coating inside the vesicles (gas bubbles) of basalt or rhyolite, rarely with any interesting features. If you look very closely, these coatings tend to have a bubbly, or somewhat botryoidal (grape-like) look, though normally they are very poorly defined. The coatings also form a base for other vesicular minerals to grow upon. Zeolites, in particular, are often found growing atop a bed of saponite.

As a clay mineral, saponite's crystal structure is mostly microscopic. It is very soft and fragile, and you're not likely to confuse it with other minerals. Chlorite forms similar coatings inside vesicles (gas bubbles), but it is harder, darker and sometimes has a more observable crystal shape. Saponite is also not very collectible or desirable by itself, but it often occurs alongside other vesicular minerals.

WHERE TO LOOK: Find saponite where you'd find basalt, especially volcanic regions such as Flagstaff and Clifton.

Orange scheelite

Pyramidal crystal

Above specimen under short-wave ultraviolet light

Scheelite

HARDNESS: 4.5–5 **STREAK:** White-yellow

Occurrence

ENVIRONMENT: Mine dumps, mountains

WHAT TO LOOK FOR: Orange pyramid-shaped crystals that glow bright white or blue under short-wave ultraviolet light

SIZE: Most scheelite occurs as very small crystals that are no larger than a pea

COLOR: Yellow to orange, flesh-colored, colorless to white

OCCURRENCE: Rare

NOTES: Scheelite is a rare tungsten-based mineral that is quite difficult to find in Arizona. Its crystals are pyramid-shaped when well-formed, but most scheelite is found as small masses, or as grains within rocks, and such scheelite is difficult to detect. And while the most highly desired scheelite is a translucent orange color, much of it will be white or colorless and a specimen can fool the casual observer into thinking it is a piece of quartz. The best identifier for scheelite is its fluorescence. Under short-wave ultraviolet light, scheelite's tungsten content causes it to glow bright blue or white. In fact, some collectors will hunt the mineral at night with an ultraviolet light in hand. Fluorite can fluoresce in similar colors, though fluorite is much more common and is slightly softer than scheelite. Scheelite also has a high specific gravity (it feels heavy for its size). Scheelite mostly occurs in metamorphic rocks where a molten granite formation contacted an existing limestone formation. It will be found alongside epidote, garnet and calcite.

WHERE TO LOOK: Scheelite is rare and not found in many locations. The mountains near Hilltop in the southeast corner of the state are a known location.

Tightly layered texture

Shiny mica schist

Mica schist

Shiny gray mica

Dark secondary mineral inclusions

Specimen courtesy of Hallie Edwardson

Schist

HARDNESS: N/A **STREAK:** N/A

ENVIRONMENT: Mountains, buttes, roadcuts, washes

Occurrence

WHAT TO LOOK FOR: Highly layered and compressed rock, sometimes very shiny or "glittery"

SIZE: Schist can be found in any size

COLOR: Varies greatly—primarily black to gray or white

OCCURRENCE: Common

NOTES: Schist, like gneiss, is the result of metamorphosis, the process of chemical transformation that occurs when a rock undergoes heat and great pressure. As a rock undergoes metamorphosis, its component minerals are reordered and form into layers. Sometimes this layering process is incomplete and some of the rock's original appearance is visible; this happens in gneiss, which is the general name for rocks with less than half of their minerals arranged into layers. Schists, on the other hand, have more than half of their minerals layered and the original rock's appearance is lost. To an extent, these two terms are interchangeable; deciding whether or not a rock is more or less than half layered can be a subjective experience.

Schists can be very compact due to their metamorphosis. Mica schists are an example of a tightly formed schist where flakes of lustrous mica minerals grouped together so closely that the rock gains a "glittery" appearance. Schists also often form new minerals within them, such as garnets and andalusite, which are generally much harder than the schist itself.

WHERE TO LOOK: Regions of metamorphic rock, such as the hills around the Prescott area, are the best places to look, though it can be found anywhere.

Glassy, flat selenite crystals

Selenite "rose"

Coating of tiny, needle-like selenite crystals

Selenite

HARDNESS: 1.5–2 **STREAK:** White

ENVIRONMENT: Desert, buttes, washes, roadcuts

WHAT TO LOOK FOR: Clear, glassy, flat crystals that are very soft

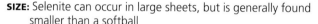
Occurrence

SIZE: Selenite can occur in large sheets, but is generally found smaller than a softball

COLOR: Colorless to white or gray, yellow to brown

OCCURRENCE: Uncommon

NOTES: One of the most collectible varieties of gypsum, selenite's glassy, clear crystals can be found all over Arizona. Most selenite occurs in sheet-like masses, as it often forms in or between beds of mud or clay material, as well as between other rocks of sedimentary origin. These flat crystals can have a layered appearance on their edges. In fact, with some specimens, you can split the crystal into two pieces by pressing a knife into the layered edge.

Some bladed selenite crystals also arrange into rosettes, or round crystal aggregates. These often occur in sandy areas, and as such the crystals themselves have sand incorporated within them, making its appearance dull and rough. The result is a "desert rose," or a selenite rosette colored by the sand it formed in. This is not to be confused with barite "desert roses," however, which are normally harder, heavier and have blunter crystal edges.

Selenite, as with all varieties of gypsum, is very soft, and can be scratched with your fingernail. Combined with its clear, flat crystals, no other tests are needed to identify it. It can occur with barite and dolomite.

WHERE TO LOOK: The bluffs in east St. David are the famous locality that has produced selenite "desert roses."

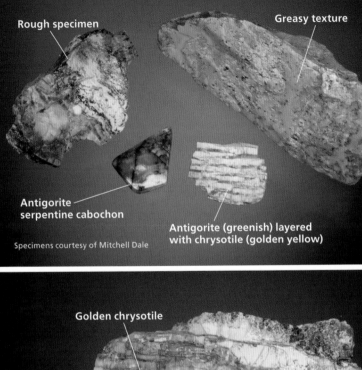

Rough specimen

Greasy texture

Antigorite
serpentine cabochon

Antigorite (greenish) layered
with chrysotile (golden yellow)

Specimens courtesy of Mitchell Dale

Golden chrysotile

Green antigorite

⚠ Serpentine Group

HARDNESS: 3–5 **STREAK:** White

ENVIRONMENT: Buttes, mountains, roadcuts, mine dumps

Occurrence

WHAT TO LOOK FOR: Greenish minerals with a greasy feel

SIZE: Serpentines are massive (they occur in compact mineral concentrations) and can be any size

COLOR: Green to yellow, golden yellow, olive green, greenish black

OCCURRENCE: Uncommon

NOTES: Serpentines are a large group of iron- and magnesium-rich minerals that all generally occur in shades of green or yellow and have a waxy or greasy feel, both of which are key characteristics. They can resemble talc, which is softer, or actinolite, which is harder. There are over twenty different serpentine minerals, though most are so similar that distinguishing them is difficult and unnecessary. The two most common types in Arizona are antigorite and chrysotile. Antigorite is a massive, often translucent serpentine that is found in shades of light to dark olive green. It exhibits a flaky, layered appearance on fresh breaks and is the most common type used in jewelry. Chrysotile is a unique fibrous variety that has a silky feel and a golden yellow sheen. The thin fibrous crystals of chrysotile occur in compact masses and are so flexible and soft that they resemble silk. Individual "threads" can be peeled off and even woven into fabrics. This flexible, fibrous form occurs in several different minerals and is known as asbestos. Asbestos minerals have become notorious for their adverse health effects, so be sure to wear a dust mask, or a respirator if you're working with large quantities.

WHERE TO LOOK: Fine examples of chrysotile have come from the Salt River Valley on the western edge of Phoenix.

Layered, sheet-like structure

Catlinite, or "pipestone"

Shale

HARDNESS: <5.5 **STREAK:** N/A

Occurrence

ENVIRONMENT: Mountains, roadcuts, washes, desert

WHAT TO LOOK FOR: Soft, layered rock, often crumbling and falling apart in sheets

SIZE: Shale occurs massively and can be any size

COLOR: White to gray, brown to black, green, red

OCCURRENCE: Common

NOTES: Shale is a sedimentary rock, meaning that it formed after material deposited by the elements hardened, forming a rock. There are many varieties of shale, each of which is determined by how lithified, or solidified, the rock is and which material hardened to form it, such as mud, clay, or silt. Most shale is quite soft and generally has a Mohs hardness of less than 5.5, which makes it weather easily. Water accelerates shale's destruction and causes the individual layers of material to separate and fall apart in sheets. The easiest way to identify shale is by its soft, layered habit (form). Not all layered rock in Arizona is shale; sandstone, which is very abundant, may also have layers. Telling the two apart is easy, as shale does not have the rough, sandy texture of sandstone. Shale occurs in many colors. It is primarily white to gray or green, but also occurs in reds or browns when iron is present. Mudstone is a particularly fine-grained variety of shale that is very soft. High quality, tightly formed red- and cream-colored mudstone is called catlinite, or, more commonly, pipestone; it has been used for carvings for centuries. Catlinite is often found in beds with quartzite, but catlinite is quite uncommon.

WHERE TO LOOK: Shale is very common; look anywhere there are rocky outcroppings or cliffs in desert areas.

Massive siderite (brown) in quartz (white)

Siderite crystal face

Massive siderite

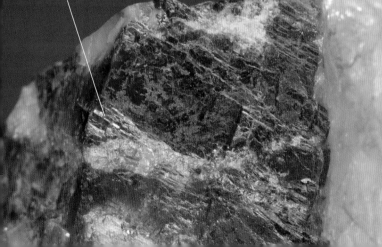

Brown, glassy crystal face

Siderite

HARDNESS: 3.5–4 **STREAK:** White-yellow

Occurrence

ENVIRONMENT: Washes, desert, mine dumps, roadcuts

WHAT TO LOOK FOR: Hexagonal, glassy brown crystals or coatings found associated with sedimentary rocks, such as shale

SIZE: Crystals are generally thumbnail-sized or smaller, while massive siderite (compact mineral concentrations) can occur in any size

COLOR: Light to dark brown or brownish red

OCCURRENCE: Uncommon

NOTES: Siderite is an iron-based mineral that is found in sedimentary rocks including shale and clay. It can form in large masses within these sedimentary beds and often forms small, brown six-sided or box-like crystals. Massive siderite appears as glassy brown coatings or veins on or in its surrounding rock. Massive forms also tend to appear layered.

Siderite occurs with a number of other minerals, such as fluorite, barite, galena, and it commonly is found alongside calcite. Of course, it also occurs with other similar iron-based minerals, especially limonite. Limonite is a name given to nondescript masses of iron oxides that are often brown or orange in color. Limonite and siderite's similar coloration might be confusing at first, but limonite never has the glassy luster of siderite. In addition, limonite has no crystal structure, whereas even massive siderite will have some crystal shape. But things are not always that simple—limonite sometimes actually replaces siderite, resulting in limonite with the exact shape and crystal structure of the siderite it replaced. In this case, streak color and luster are good tests.

WHERE TO LOOK: Siderite most often forms in sedimentary deposits with shales and clays and can be found alongside those minerals.

Quartz

Acanthite crystal coating (black)

Silver wire

Silver nugget

Silver ore

Silver

HARDNESS: 2.5–3 **STREAK:** Silver-gray to white

ENVIRONMENT: Mine dumps

WHAT TO LOOK FOR: Silver-colored metal, often blacked with tarnish

Occurrence

SIZE: Silver can be found golf ball-sized, but it is more commonly pea-sized or smaller

COLOR: Silver when fresh, black to gray when tarnished

OCCURRENCE: Very rare

NOTES: Like copper, silver is one of the minerals that has been mined in Arizona for decades, though actual silver specimens that exhibit massive chunks or crystals are extremely scarce in Arizona. Most of the silver being mined is in the form of ore, rocks rich in silver that can be processed and heated to extract the metal. Most of this ore is unattractive and you cannot visually determine that the rock contains silver at all. A very rare variety of silver crystal is silver wire. Silver wire is a natural occurrence that appears exactly how it sounds—as a long, thin strand of silver that tapers to a point. "Standard" silver crystals often appear dendritic, or tree-like, in nature. Their branching growths look more like a plant than a metal.

You're very unlikely to confuse silver with anything else. Its bright silver color is normally always enough to tell it apart, though its hardness and flexibility are also good identifiers. When tarnished, silver turns gray or black. This dark color is actually another mineral called acanthite which can form its own crystals on top of silver.

WHERE TO LOOK: Mine dumps in the Tombstone Hills, west of Tombstone, and near Hilltop, southeast of Willcox, as well as many other mine dumps in southeastern Arizona.

Specimens of botryoidal (grape-like) smithsonite

Smithsonite lining a cavity

Smithsonite

HARDNESS: 4–4.5 **STREAK:** White

ENVIRONMENT: Mine dumps

WHAT TO LOOK FOR: Green or blue rounded, botryoidal crusts on host rock

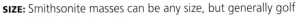
Occurrence

SIZE: Smithsonite masses can be any size, but generally golf ball-sized and smaller

COLOR: Green to blue, yellow, white or gray

OCCURRENCE: Uncommon

NOTES: Smithsonite is a very collectible zinc-based mineral that mostly occurs in shades of green in Arizona, but it can also be blue or yellow, though this is rare. The mineral forms botryoidal (grape-like) crusts or stalactitic masses that have a rounded, pillowy-looking texture. Few other minerals in Arizona occur this way and smithsonite is generally easy to identify because of it. Hemimorphite can form rounded masses similar to smithsonite, but hemimorphite is slightly harder, doesn't normally occur in shades of green, and often has larger, more visible crystals. Prehnite, a mineral rare in Arizona but very common in other states, can have the same crystal appearance and color but is much harder than smithsonite.

Smithsonite occurs with copper- and lead-based minerals, such as malachite, azurite and galena, and is best formed in dry climates rich with limestone. Smithsonite is named in honor of James Smithson who founded the Smithsonian Institution in Washington, DC.

WHERE TO LOOK: The mine dumps in the Santa Rita Mountains, Patagonia Mountains and Empire Mountains, all northeast of Nogales and southeast of Tucson, have yielded smithsonite.

Dark yellow crystal edges

Black coloration due to the presence of iron

Tetragonal (pyramid-like) crystals

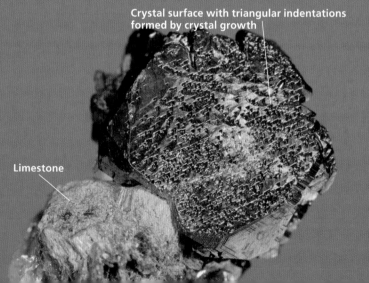

Crystal surface with triangular indentations formed by crystal growth

Limestone

Sphalerite

HARDNESS: 3.5–4 **STREAK:** Light brown

Occurrence

ENVIRONMENT: Mine dumps, mountains

WHAT TO LOOK FOR: Small, dark crystals with many faces

SIZE: Individual crystals are generally thumbnail-sized or smaller

COLOR: Translucent, dark yellow-green to brown, also red or black

OCCURRENCE: Uncommon

NOTES: Sphalerite is the world's primary ore of zinc, as it consists of a simple chemical mixture of zinc and sulfur. Its crystals are tetragonal (pyramid-shaped) and are often twinned (two or more crystals occur together). These crystals are primarily green, dark yellow or brown but are more opaque and dark (even black) when iron is present in its composition. Sphalerite crystals often occur in masses with each other and sometimes have distorted faces that appear curved. These misshapen crystals can make the mineral harder to identify, but its streak and hardness are the primary identifiers.

Sphalerite is generally found in limestone deposits and is closely associated with galena, many times occurring in the same specimen. Distinguishing the two can be difficult when sphalerite is rich in iron impurities and becomes black and opaque. The easiest way to tell them apart is by their streak colors. Sphalerite's streak will be lighter and browner than galena's. When sphalerite is altered, or changed by other minerals entering its composition, it can change into one of Arizona's other zinc-based minerals, such as hemimorphite, smithsonite or willemite.

WHERE TO LOOK: Sphalerite can be found in the hillsides and mine dumps in the Chiricahua Mountains near, but not in, the Chiricahua National Monument.

Soft, soap-like texture

Green coloration

Talc

HARDNESS: 1 **STREAK:** White

ENVIRONMENT: Roadcuts, mountains, buttes

WHAT TO LOOK FOR: Extremely soft, greenish masses that feel soapy

Occurrence

SIZE: Talc occurs massively (in compact mineral concentrations) and can be found in any size

COLOR: Apple-green to white

OCCURRENCE: Common

NOTES: Talc is one of the softest minerals on the planet and is the softest mineral on the Mohs hardness scale, with a hardness of 1. In fact, it is so soft that your fingernail will sink deeply into its surface. Crystals are rare and talc almost always occurs as compact masses that have a scaly, foliated (layered) surface. While most talc is white, finer quality talc is green. Talc is generally opaque, but very pure and well-formed varieties can be translucent and quite attractive for collectors.

Talc is often referred to as "soapstone" because of its greasy or waxy feel, much like a bar of soap. This fact, combined with its incredible lack of hardness, differentiate it from actinolite, a green variety of the mineral tremolite, with which it frequently occurs. Another mineral called magnesite could be confused with talc, but magnesite is much harder and is generally yellow in color. Generally speaking, it is very hard to confuse talc with another mineral.

Talc has been carved into tools and decorations for centuries because of the little effort it takes to shape it.

WHERE TO LOOK: The Patagonia Mountains, east of Nogales, and the Little Dragoon Mountains, northeast of Benson, are good places to start looking.

Malachite (green)   Tenorite matrix (brown)

Dark, massive, featureless structure

Slice of malachite (green) and chrysocolla (blue) containing tenorite (black)

Tenorite

HARDNESS: 3.5–4 **STREAK:** Black

Occurrence

ENVIRONMENT: Mine dumps, mountains, road-cuts, desert

WHAT TO LOOK FOR: Black, dull, featureless masses or bands occurring with copper-based minerals

SIZE: Tenorite occurs massively and can be found in any size

COLOR: Black to blackish brown or gray

OCCURRENCE: Common

NOTES: Tenorite is a general name for a mineral that consists entirely of oxidized copper, just as limonite is made up of oxidized iron. As such, the name tenorite can be considered a catch-all term for dark, unidentified copper oxides. Tenorite crystals do exist but are extremely rare and what you'll most likely find are nearly featureless masses with a very dull luster. It is a frequent occurrence in copper deposits, but mostly as black bands or masses in or on other copper materials like malachite and chrysocolla. Tenorite can also form in large masses which act as a matrix for the more collectible blue and green copper minerals.

Tenorite can be confused with several other dark-colored minerals, such as hematite or pyrolusite. Hematite has a reddish streak, is harder and usually has a metallic luster, unlike tenorite. Pyrolusite often has a fibrous structure and higher luster but is also harder. And, of course, if the minerals occurring with the specimen in question are copper-related, your specimen is probably tenorite.

WHERE TO LOOK: Nearly every copper mine dump will yield tenorite, but you will also find it in more natural conditions near copper deposits. The southeast corner of the state is a good place to look.

Tiny, square crystals

⚠️ **Torbernite**

HARDNESS: 2–2.5   **STREAK:** Pale green

ENVIRONMENT: Mine dumps

WHAT TO LOOK FOR: Square, green crystals that are often very small and thin

Occurrence

SIZE: Torbernite crystals can vary in size, but many are tiny and measure less than one millimeter across

COLOR: Deep green to yellow-green

OCCURRENCE: Rare

NOTES: When uranium, a radioactive element, is present in a mineral, it often turns it one of two colors: bright yellow or green. Torbernite is no exception, and it contains more than fifty percent uranium. Its radioactivity alone should make you wary of it and avoid handling it, but to make matters worse, it often dehydrates when kept in collections. When this happens, the crystals break down and form a dust that is very harmful if inhaled. So not only will you lose your specimen, it can create a hazard for anyone who comes into contact with it. It is for these reasons that it is not advised to collect torbernite.

Luckily, Arizona's torbernite is generally very small in size, with a majority of the crystals measuring less than a millimeter across. At this size, it is relatively safe to handle for short amounts of time. Torbernite is quite easy to identify because its crystals form perfect, flat squares, or series of intergrown squares. This crystal habit (form), combined with its color, is rare in any other mineral in Arizona. Torbernite generally occurs in veins within granite or granitic pegmatites (the lowest portion of a granite formation).

WHERE TO LOOK: The Sierra Ancha Mountains have produced torbernite and are located northeast of Phoenix and southeast of Payson.

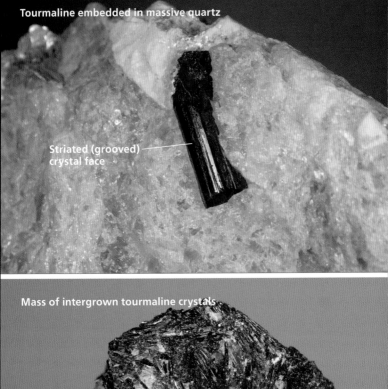

Tourmaline embedded in massive quartz

Striated (grooved) crystal face

Mass of intergrown tourmaline crystals

Tourmaline Group

HARDNESS: 7–7.5 **STREAK:** White

Occurrence

ENVIRONMENT: Mountains, roadcuts, mine dumps, buttes

WHAT TO LOOK FOR: Black, grooved crystals that are long and thin

SIZE: Tourmaline crystals are always much thinner than they are long and sometimes are several inches long but only a millimeter or two wide

COLOR: Black, brown and very rarely pink or green

OCCURRENCE: Uncommon

NOTES: The name tourmaline actually refers to a group of closely related minerals that exhibit the same crystal structure. These crystals are generally much longer than they are wide, with striated (grooved) crystal faces and a cross-section that resembles a "bulging triangle," or a triangle with rounded sides. While vividly colored varieties exist, you'll only find the black variety, called schorl, in Arizona. Schorl occurs in several mineral environments. The finest, largest schorl crystals are found in granite pegmatites where the tourmaline was allowed a great amount of time to crystallize. It is also found within granite rocks as small, poorly formed grains and as short crystals embedded in metamorphic rock, forming much the way garnets do. Individual crystals are most desirable to collectors but less common than small intergrown crystals which often appear as radiating, fan-like aggregates. Tourmaline is also often found embedded within quartz and makes for attractive, high contrast specimens. Few minerals are easy to confuse with tourmaline, as its hardness, distinct crystal habit (form) and growth environments are normally enough to identify it. Tourmaline is commonly found with quartz, micas and feldspars.

WHERE TO LOOK: The Bradshaw Mountains, south of Prescott, and the White Picacho Mountain district, northwest of Phoenix.

219

Tremolite

Light-colored, radiating, fibrous structure

Actinolite

Dark, fibrous structure

⚠ Tremolite

HARDNESS: 5–6 **STREAK:** White

ENVIRONMENT: Mine dumps, mountains, desert, buttes

Occurrence

WHAT TO LOOK FOR: Light-colored, fibrous mineral found in limestone-rich areas

SIZE: Tremolite formations are generally small and no bigger than a golf ball

COLOR: White to cream-colored, gray, yellow

OCCURRENCE: Uncommon

NOTES: Tremolite is a fibrous mineral that is rich in magnesium and is found in marble (metamorphosed limestone) deposits. Tremolite also occurs in schists and gneisses with serpentine, talc and calcite. Its crystal fibers can be extremely thin and small and are actually a variety of asbestos. As such, tremolite can be dangerous—if these particles become airborne, they can be very harmful if inhaled. In fact, tremolite asbestos is considered more damaging to human health than the more common serpentine asbestos. Always handle this mineral with care and wear a respirator if you wish to hunt for or dig tremolite.

While magnesium is the primary ingredient in tremolite, iron can also be present. As the level of iron increases, the mineral becomes darker and greener. Iron-rich tremolite is called actinolite and is a dark, rich green. In fact, most specimens of jade, the gemstone made famous by Asian culture, are actually high quality actinolite. And when tremolite undergoes metamorphosis, it becomes diopside.

WHERE TO LOOK: The Sierrita Mountains mine dumps, south of Tucson, have produced a lot of tremolite.

Tuff

HARDNESS: <5 **STREAK:** N/A

ENVIRONMENT: Desert, buttes, washes, roadcuts

Occurrence

WHAT TO LOOK FOR: Light-colored rock consisting of compacted fragments of other rocks

SIZE: Tuff can occur in any size, from pebbles to boulders

COLOR: White to pale brown, gray or yellow

OCCURRENCE: Very common

NOTES: Tuff forms as a result of volcanic activity. Ancient volcanoes erupted and threw out tons of ash and pulverized rock, which fell to the earth and settled into beds. These beds then lithified (compressed) into tuff. There are different varieties of tuff and each depends on the type of material contained within it. Rhyolitic tuff, for example, primarily contains rhyolite ash. Many tuffs exhibit layers made by different volcanic ash deposits.

Even the hardest tuff can be scratched with a knife, while you can pick apart the softer varieties with your bare hands. Tuff's hardness depends on how welded, or solidified, the tuff is. Thicker tuff deposits tend to be more compacted while thinner beds of tuff are less compacted. Some tuff contains larger fragments mixed within the ash, and when the material is compacted and hardened these larger fragments help hold the tuff together.

Tuffs are generally fairly easy to identify, as all tuffs tend to have a gritty, rough feel. To locate tuffs, look for a fine-grained rock with slightly larger fragments mixed throughout. Fine, poorly welded tuffs can be powdery and are easy to break apart.

WHERE TO LOOK: Tuff is easily found in most sedimentary desert areas.

Mimetite (orange)

Thin turquoise coating

Turquoise vein

Turquoise filling a cavity

Broken piece of massive turquoise

Polished varieties of turquoise

Turquoise

HARDNESS: 5–6 **STREAK:** Pale green

ENVIRONMENT: Mine dumps, mountains, roadcuts

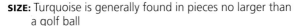

Occurrence

WHAT TO LOOK FOR: Hard blue masses filling cracks and cavities in rock

SIZE: Turquoise is generally found in pieces no larger than a golf ball

COLOR: Deep to pale blue, bluish green or pale green

OCCURRENCE: Uncommon

NOTES: If one were to choose Arizona's most well-known and desirable collectible mineral, it would have to be turquoise. Collectors covet its sky-blue color and it has been popular for use in jewelry for decades. Turquoise crystals are so rare that most collectors don't even know that they can exist. Instead, turquoise often occurs as compact seams filling cracks within broken brown or white rocks. This massive form of turquoise, while hard, is often unstable and crumbles easily. Because of this, only the finest, most solid specimens will take a good polish. Turquoise, like many of Arizona's blue or green minerals, is colored by copper. Of these minerals, only chrysocolla is similar enough in color and structure to be confused with turquoise. Chrysocolla, however, is much softer unless you have a piece of "gem chrysocolla." This popular gem variety is actually quartz colored by chrysocolla inclusions, so its hardness will be that of quartz, which is much harder than turquoise. Once you have determined that a specimen is not chrysocolla, turquoise's distinct color is normally enough to identify it.

WHERE TO LOOK: The Bisbee area mines are famous for their richly colored turquoise. Also, the Silver Bell Mountains, west of Tucson, have many localities, as well as many other locations throughout the state.

Surface coating of vanadinite (red)

Vanadinite (orange) on quartz (white)

Large "hopper" crystal

Hexagonal crystals

Vanadinite

HARDNESS: 3 **STREAK:** White to pale yellow

Occurrence

ENVIRONMENT: Mine dumps

WHAT TO LOOK FOR: Small, hexagonal, barrel-shaped crystals coating the surface of rock

SIZE: Vanadinite crystals are small and most are less than a quarter of an inch in length

COLOR: Red, orange, brown or yellow

OCCURRENCE: Uncommon

NOTES: Vanadinite is named for the element vanadium, which is one component of the mineral. However, it is primarily a lead-based mineral and like other lead minerals such as wulfenite, it is brightly colored and has a high luster. Vanadinite's unique little crystals are hexagonal (six-sided) barrels that often lie in no particular direction on the surface of another mineral or rock. These crystals are also sometimes hollow, and with a good magnifying glass you can see a small hole at the center of their end faces. Rarely, a vanadinite crystal will grow tall and slender, widening at its top, and the sides of the crystal will grow taller than the center forming a "cup." These are called "hopper" crystals; they can be over an inch long and are very rare in Arizona. As a lead mineral, vanadinite frequently occurs with galena and wulfenite as well as barite and quartz. Of these, the only mineral with which it could be confused is wulfenite, due to their similar colors and luster. Since their hardness is the same, you'll have to rely on crystal structure to tell the two apart. Wulfenite's crystals are flat squares that normally occur in shades of yellow or orange.

WHERE TO LOOK: Mines in the Cerbat Mountains, northwest of Kingman, and the Patagonia Mountains, northeast of Nogales, are known for their vanadinite specimens.

Green willemite crystals

Mimetite (orange)

Willemite in marble

Above specimen under short-wave ultraviolet light

Willemite

HARDNESS: 5.5 **STREAK:** White

ENVIRONMENT: Mine dumps, mountains

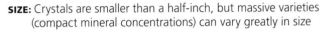

Occurrence

WHAT TO LOOK FOR: Ideal specimens have small, hexagonal, green crystals, but occurs more often massively in marbles

SIZE: Crystals are smaller than a half-inch, but massive varieties (compact mineral concentrations) can vary greatly in size

COLOR: Yellow to green, colorless to white

OCCURRENCE: Rare

NOTES: Some varieties of willemite are highly fluorescent and glow brightly under short-wave ultraviolet light, which makes it a favorite among collectors. Unfortunately, much of Arizona's willemite is not fluorescent because of impurities—though a few localities do produce the fluorescent variety. As a pure, well-formed mineral, willemite is found as small hexagonal (six-sided) prisms, though these are rare. Willemite is frequently found as a thin coating on top of other rocks and minerals, though it's most commonly found within rock, especially marble, in which it is nearly impossible to see with the naked eye. In such cases, an ultraviolet light is necessary and causes willemite to glow if it is of the rare fluorescent variety. Since crystals are scarce and willemite most commonly occurs as nondescript portions of rock, identification is difficult when specimens aren't fluorescent. Hardness can sometimes help identify it, as willemite is harder than the marbles in which it frequently occurs. The best thing to do is hunt in an area known for willemite and then ask an expert about your finds.

WHERE TO LOOK: Look in the mine dumps in the Globe Hills, north of Globe, the Huachuca Mountains and near Hilltop.

Intergrown crystals

Square, glassy crystals

A rare crystal

Wulfenite crystals (yellow) on the surface of a rock

Wulfenite

HARDNESS: 3 **STREAK:** White

ENVIRONMENT: Mine dumps

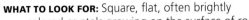

WHAT TO LOOK FOR: Square, flat, often brightly colored crystals growing on the surface of rocks or other crystals

Occurrence

SIZE: Wulfenite crystals are small and don't normally grow larger than your thumbnail

COLOR: Yellow to orange, red, or brown

OCCURRENCE: Uncommon

NOTES: One of Arizona's best-known and collectible crystals, wulfenite is a must-have for any collection. Wulfenite's crystals are flat, plate-like rectangles with tapered edges that are vividly colored in reds, oranges and yellows. They are often found standing up off of the surface of other minerals. Combined with its brilliant luster, these traits make wulfenite difficult to confuse with other minerals. In addition, wulfenite almost never occurs massively, as most other minerals will. Most wulfenite crystals are pea-sized and smaller, but they can grow to thumbnail size. Sometimes wulfenites will form tightly together and create a group of tangled, intergrown crystals. While interesting, these aggregates don't often contain well-formed individual crystals and therefore are less collectible. The Red Cloud Mine in Yuma county is one of Arizona's wulfenite localities and is renowned as one of the world's finest sites for wulfenite. This mine has produced rare wulfenite crystals that are unusually bright red and orange because of inclusions of the element chromium.

WHERE TO LOOK: Wulfenite is prevalent in many Arizona mines, including those in the Patagonia Mountains, the Empire Mountains, the White Picacho Mountains and the Trigo Mountains.

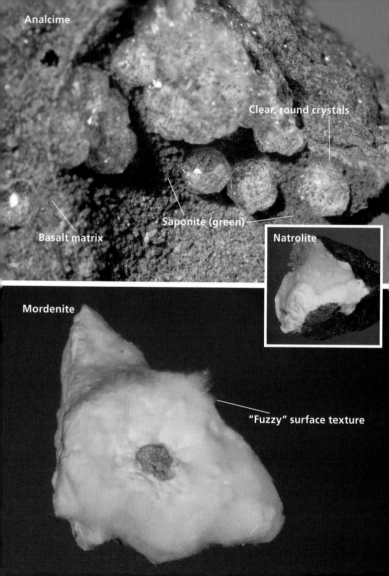

Analcime

Clear, round crystals

Saponite (green)

Basalt matrix

Natrolite

Mordenite

"Fuzzy" surface texture

Zeolite Group

Occurrence

HARDNESS: 3.5–5.5 **STREAK:** Colorless to white

ENVIRONMENT: Mountains, mine dumps, buttes, desert

WHAT TO LOOK FOR: Light-colored minerals growing within cavities in rock

SIZE: Most zeolites have pea-sized or smaller crystals

COLOR: Colorless, white to gray, reddish to pink

OCCURRENCE: Common

NOTES: Zeolites encompass a large and diverse group of minerals comprised of aluminum, sodium and silica (quartz). Most zeolites are light-colored and are found in shades of white, gray or pink. Analcime, one of Arizona's more common zeolites, can be white, but is most often colorless and forms as tiny ball-like crystals. Natrolite, another zeolite, can be colorless, but is often white or pink and forms fine, needle-like crystals that radiate outward from a central point. Stilbite is also fairly common and forms pink or red bow tie-shaped crystals that are striated (grooved). Mordenite is a rare variety that forms white "hairs" that are so tiny and thin that the specimen appears "fuzzy." Laumontite is a pink to orange variety that dehydrates and crumbles easily. Most zeolites aren't easily confused with other minerals. Calcite has a similar hardness and color, but doesn't crystallize in the same way. In many places around the world, zeolites can be free-growing and become quite large. In most of the U.S. (including Arizona), zeolites are found in vesicles (gas bubbles) within basalt or rhyolite as tiny, well-formed crystals. Most zeolites are soft and brittle; don't try to separate the crystals or you'll probably destroy the specimen.

WHERE TO LOOK: Many zeolites are found in the Horseshoe Dam area, north of Phoenix.

Glossary

AGGREGATE: An accumulation or mass of crystals

ALTER: Chemical changes within a rock or mineral due to the addition of mineral solutions

AMYGDULE: A vesicle, or gas bubble, filled with a secondary mineral

ASBESTOS: A very fibrous, flexible, silky mineral formation

ASSOCIATED: Minerals that often occur together due to similar chemical traits

BAND: An easily identified layer within a mineral

BED: A flat mass of rock, generally sedimentary

BOTRYOIDAL: Rounded masses resembling a bunch of grapes

BRECCIA: A coarse-grained rock composed of broken, angular rock fragments solidified together

CHALCEDONY: A massive, microcrystalline variety of quartz

CLEAVAGE: The way a mineral breaks along the planes of its crystal structure, which reflects its internal crystal shape

COMPACT: Dense, tightly formed rocks or minerals

CONCENTRIC: Circular, ringed banding, like a bull's-eye pattern, with larger rings encompassing smaller rings

CRYSTAL: A solid body, with a repeating atomic structure, consisting of an element or chemical compound

CUBIC: Box-like structure where all sides are of equal size

DESICCATE: Complete dehydration of a mineral

DRUSY: A coating of small crystals on the surface of another rock or mineral

EARTHY: Resembling soil; dull luster and rough texture

EFFERVESCE: When a mineral placed in an acid gives off bubbles

FELDSPAR: An extremely common and diverse group of light-colored minerals that are most prevalent within rocks and make up the majority of the earth's crust

FIBROUS: Fine, rod-like crystals that resemble cloth fibers

FLUORESCENCE: The property of a mineral to give off visible light when exposed to ultraviolet light radiation

GLOBULAR: Crystal formation resembling a ball, or globe

GNEISS: A rock that has been metamorphosed so that some of its minerals are aligned in parallel bands

GRANITIC: Pertaining to granite or granite-like rocks

GRANULAR: A texture or appearance of rocks or minerals that consist of grains or particles

HEXAGONAL: A six-sided structure

HOPPER: A crystal with sides taller than its center

HOST: A rock or mineral on or in which other rocks and minerals occur

IGNEOUS ROCK: Rock resulting from the solidification of molten rock material, such as magma or lava

IMPREGNATE: When a rock or mineral is infused within another mineral

IMPURITY: A foreign mineral within a host mineral that often changes the color of the host

INCLUSION: A mineral, often well-crystallized, that is encased or impressed into a host mineral

IRIDESCENCE: When a mineral exhibits a rainbow-like play of color

LAMELLAR: Composed of thin crystals arranged in book- or gill-like aggregates

LAVA: Molten rock that has reached earth's surface

LITHIFICATION: When material is subjected to pressure which forces liquids out of it and causes it to harden into a rock

LUSTER: The way in which a mineral reflects light off of its surface, described by its intensity

MAGMA: Molten rock that remains deep in the earth

MASSIVE: Rocks that consist of generally the same texture; minerals that don't occur in individual crystals but rather as a solid, compact concentration

MATRIX: The rock in which a mineral forms

METAMORPHIC ROCK: Rock derived from the altering of existing igneous or sedimentary rock through the forces of heat and pressure

METAMORPHOSED: A rock that has already undergone metamorphosis

MICA: A group of minerals that occur in thin flakes arranged in book-like aggregates

MICACEOUS: Mica-like in nature; a mineral consisting of thin, flexible sheets

MICROCRYSTALLINE: Crystal structure too small to see with the naked eye

MINERAL: A naturally occurring chemical compound or native element that solidifies with a definite internal crystal structure

NODULE: A rounded mass consisting of a mineral

OCTAGONAL: An eight-sided structure

OCTAHEDRAL: A structure with eight-faces, resembling two pyramids placed base-to-base

OPAQUE: Material that lets no light through

ORE: Material from which metals can be extracted

OXIDATION: The process of a metal combining with oxygen, which can produce new colors or minerals

OXIDE: A mineral formed from the linking of oxygen with a metallic element

PEGMATITE: The lowest portion of a granite formation, where the minerals within the magma are allowed great amounts

of time to cool and therefore fully crystallize, often resulting in very large, and sometimes rare, crystals

PHENOCRYST: A crystal embedded with igneous rock that
solidified before the rest of the surrounding rock, thus
retaining its true crystal shape

PLATY: A formation of a thin, flat nature; plate-like

PORPHYRY: An igneous rock containing many phenocrysts

PRISMATIC: Crystals with length greater than their width

PSEUDOMORPH: When one mineral replaces another, keeping
the outward appearance of the initial mineral

PYRAMIDAL: Crystals resembling a pyramid with four or more
total faces

PYROXENE: A group of dark (green to black), rock-building
minerals that make up many dark-colored rocks like
basalt or gabbro

RADIATING: Circular crystals or crystal aggregates growing
outward from a central point

ROCK: An massive aggregate of minerals

ROCK-FORMING: Refers to a mineral important in the creation
of rocks

ROSETTE: Aggregates of thin crystals arranged in ball-like
formations, which resemble the petals of a flower

SCHIST: A rock that has been metamorphosed so that most of
its minerals have been arranged into parallel bands or layers

SCORIA: Basalt with a very high number of vesicles (cavities caused by gas bubbles)

SECONDARY: A rock or mineral that formed later than the rock surrounding it

SEDIMENT: Fine particles of rocks or minerals deposited by water or wind, e.g. sand

SEDIMENTARY ROCK: Rock derived from sediment being cemented together

SILICA: Silicon dioxide, more commonly known as quartz

SPECIFIC GRAVITY: The ratio of the density of a given solid or liquid to the density of water when the same amount of each is used, e.g. the specific gravity of galena is approximately 7.5, meaning a sample of galena is about 7.5 times heavier than the same amount of water

SPECIMEN: A sample of a rock or mineral

STALACTITIC: Resembling a stalactite, which is a cone-shaped mineral deposit

STRIATED: Parallel grooves in the surface of a mineral

TABULAR: A rock or mineral structure in which one dimension is notably shorter than the other two, often resulting in flat, plate-like shapes

TETRAGONAL: A crystal structure where one dimension is longer than the other two (which are equal), resulting in crystals that resemble two tall pyramids placed base-to-base

TRANSLUCENT: A material that lets some light through

TRANSPARENT: A material that lets enough light through as to be able to see what lies on the other side

TRAP ROCK: Any dark rock, such as basalt, that can trap other minerals within its vesicles

TWIN: An intergrowth of two or more crystals

VEIN: A mineral which has filled a crack or similar opening in a host rock

VESICLE/VESICULAR: A cavity created in an igneous rock by a gas bubble trapped when the rock solidified

VUG: A small cavity within a rock or mineral that is generally lined with different mineral crystals

WELDING/WELDED: When sediments, such as hot volcanic ash, are compacted and hardened due to cooling and pressure

Arizona Rock Shops and Museums

ARIZONA DISCOVERIES
317 Main Street
Jerome, AZ 86331
(928) 634-5716

ARIZONA MINING AND MINERAL MUSEUM
1502 West Washington
Phoenix, AZ 85007
(602) 721-1611

ARIZONA-SONORA DESERT MUSEUM
2021 North Kinney Road
Tucson, AZ 85743
(520) 883-2702

BISBEE MINING AND HISTORICAL MUSEUM
5 Copper Queen Plaza
Bisbee, AZ 85603
(520) 432-7071
www.bisbeemuseum.org

BLUE OPAL ART GALLERY AND JAY-R MINE
2234 N. Evans Road
Huachuca City, AZ 85616
(520) 456-9202

COPPER CITY ROCK SHOP
566 E. Ash Street, Highway 60-70
Globe, AZ 85501
(928) 425-7885

DAVID SHANNON MINERALS
6649 East Rustic Drive
Mesa, AZ 85215
(480) 985-0557 (Call ahead!)

JIM GRAY'S PETRIFIED WOOD COMPANY
On the corner of Highways 77 and 180
Holbrook, AZ
(928) 524-1842

THE ROCK PILE
250 West Main Street
Quartzsite, AZ 85346
(928) 575-6108

Bibliography and Recommended Reading

Books about Arizona Rocks and Minerals

Anthony, John W., et al., *Mineralogy of Arizona, 3rd Edition*. Tucson: The University of Arizona Press–Tucson, 1995.

Blair, Gerry, *Rockhounding Arizona*. Guilford: Falcon Press Publishing Company, 1992.

Bearce, Neil R., *Minerals, Fossils, and Fluorescents of Arizona*. Tempe: Arizona Desert Ice Press, 2006.

General Reading

Bates, Robert L., editor, *Dictionary of Geological Terms, 3rd Edition*. New York: Anchor Books, 1984.

Chesteman, Charles W., *The Audubon Society Field Guide to North American Rocks and Minerals*. New York: Knopf, 1979.

Mottana, Annibale, *et al., Simon and Schuster's Guide to Rocks and Minerals*. New York: Simon and Schuster, 1978.

Pellant, Chris, *Rocks and Minerals*. New York: Dorling Kindersley Publishing, 2002.

Pough, Frederick H., *Rocks and Minerals*. Boston: Houghton Mifflin, 1988.

Index

About the Authors

Bob Lynch is a lapidary and jeweler living and working in Two Harbors, Minnesota. He has been cutting and polishing rocks and minerals since 1973, when he desired more variation in gemstones for his work with jewelry. When he moved from Douglas, Arizona, to Two Harbors in 1982, his eyes were opened to Lake Superior's entirely new world of minerals. In 1992, Bob and his wife Nancy, whom he taught the art of jewelry making, acquired Agate City Rock Shop, a family business founded by Nancy's grandfather, Art Rafn, in 1962. Since the shop's revitalization, Bob has made a name for himself as a highly acclaimed agate polisher and as an expert resource for curious collectors seeking advice. Now, the two jewelers keep Agate City Rocks and Gifts open year-round and are the leading source for Lake Superior agates, with more on display and for sale than any other shop in the country.

Dan R. Lynch has a degree in graphic design with emphasis on photography from the University of Minnesota Duluth. But before his love of the arts came a passion for rocks and minerals, developed during his lifetime growing up in his parents' rock shop. Combining the two aspects of his life seemed a natural choice and he enjoys both writing about and taking photographs of minerals. Working with his father, Bob Lynch, a respected veteran of the rock-collecting community, Dan spearheads their series of rock and mineral field guides—definitive guidebooks useful for rock hounds of any skill level. Dan takes special care to ensure that his photographs complement the text and represent each rock or mineral exactly as you'll find them. Encouraged by his wife, Julie, he works as a writer and photographer.

Notes

Notes

Notes